william sinnema

ELECTRONIC TRANSMISSION TECHNOLOGY

lines, waves, and antennas

PRENTICE-HALL, INC., *Englewood Cliffs, New Jersey* 07632

Library of Congress Cataloging in Publication Data

Sinnema, William (date)
    Electronic transmission technology.

    Includes index.
    1. Microwave transmission lines.   2. Wave guides.
3. Antennas (Electronics)   I. Title.
TK7876.S573   621.38′043   78-12568
ISBN   0-13-252221-7

Editorial/production supervision and interior design
        by Barbara Cassel
Cover design by Suzanne Behnke
Manufacturing buyer: Gordon Osbourne

Printed in the United States of America

10   9   8   7   6   5

PRENTICE-HALL INTERNATIONAL, INC., *London*
PRENTICE-HALL OF AUSTRALIA PTY. LIMITED, *Sydney*
PRENTICE-HALL OF CANADA, LTD., *Toronto*
PRENTICE-HALL OF INDIA PRIVATE LIMITED, *New Delhi*
PRENTICE-HALL OF JAPAN, INC., *Tokyo*
PRENTICE-HALL OF SOUTHEAST ASIA PTE. LTD., *Singapore*
WHITEHALL BOOKS LIMITED, *Wellington, New Zealand*

CONTENTS

8   RADIO-WAVE PROPAGATION   172

9   ANTENNA FUNDAMENTALS   201

PREFACE

The purpose of this textbook is to provide a strong background to technologists in the fundamentals of traveling waves on guided structures and in free space. A student using this book should have a knowedge of algebra and circuit theory; an acquaintance with differential equations would be helpful although not essential.

The text initially describes the physical configurations of various types of transmission lines and gives a clear picture by way of examples of how voltage or current transients behave on such lines. This greatly stimulates student interest in the phenomenon of traveling waves and gives them a good physical understanding for their behavior on the structures discussed in the rest of the text. The steady-state solutions on a uniform transmission line are fully developed mathematically and are later applied by analogy to hollow guides and free space transmission. Some of these derivations in Chapter Three may be skipped over, but the conclusions are significant and should be thoroughly understood.

The Smith Chart and applications to both lossless and lossy lines are covered in Chapter Four. The following two chapters are devoted to the impedance measurement and the matching techniques that are employed at high frequencies.

Plane waves and moding in hollow waveguides are covered in Chapter Seven. This chapter also includes the promising low loss helical waveguide and optical fibers.

Chapter Eight, in mainly a descriptive manner, discusses the various types of radio wave propagation used in radio and microwave systems. The final chapter is devoted to antennas and concludes with the design of a Master Antenna Television System.

I wish to express my appreciation to many of my associates and students for helpful suggestions. In particular I must thank Mr. Ben Wentzell for his encouragement in the completion of this text.

WILLIAM SINNEMA

one

CHARACTERISTICS OF STANDARD TRANSMISSION LINES

1-1 INTRODUCTION

The central purpose of this text is to communicate an understanding of the traveling-wave phenomenon. Although ignored in some applications, a finite time always elapses before a change at one point reaches another point. In basic circuit theory we neglect the effects of the finite time of transit of changes in current and voltage and the finite distances over which these changes occur. We assume that changes occur simultaneously at all points in the circuits. But there are situations in which we must consider the finite time it takes an electrical wave to travel and the distance it will travel. It is in these situations that we must employ traveling-wave theory.

Although this text will deal primarily with electric waves, it may be worthwhile to note some other types of waves. Sound waves in air, for instance, form a "longitudinal" wave where the particle motion (air) is in the direction of wave travel. It forms rarefractions and condensations of the air particles, which propagate at a velocity of 330 m/s at standard temperature (0°C) and pressure (14.7 lb/in.²). This velocity increases with an increase in the air pressure or with a reduction of the mass density of the medium.

Heat conduction in solids also propagates as a wave. Thermal waves correspond to a very lossy type of transmission line, and therefore experience a high attenuation. The velocity of thermal propagation increases with the conductivity but decreases with the specific heat and mass density of the medium. For many other examples one may wish to refer to the book by R. K. Moore.[1]

Traveling-wave concepts must be used whenever the distance is so great or the frequency so high that it takes an appreciable portion of a cycle for the wave to travel the distance. The meaning of "appreciable portion of a cycle" depends upon the application. In some devices, phase shifts of less than a degree (0.3% of a period) are significant. In other cases, a quarter-cycle delay may be permitted.

To obtain a feeling of where or when traveling-wave theory should be employed, we should obtain an expression for the wavelength (λ) of a wave. For sinusoidal signals we define a *wavelength* as that distance which a wave travels in one cycle, or period. In free space electric waves travel at the velocity of light, c, which is equal to

$$c = 3 \times 10^8 \text{ m/s}$$
$$= 186,000 \text{ mi/s}$$

In one period, therefore, the distance a wave travels (1 wavelength) is

$$\lambda = \text{velocity} \times \text{period}$$
$$= v \times T$$
$$= v \times \frac{1}{f} \tag{1-1}$$

where f is the frequency of the signal. In free space the wavelength will be equal to

$$\lambda = \frac{c}{f} \tag{1-2}$$

Wavelengths in free space can now be found for some typical frequencies. Some of these are tabulated in Table 1-1. If we assume that the traveling-wave technique must be employed for distances greater than $\frac{1}{10}$ wavelength, a distance of 3 mm at 10 GHz would require the use of this technique, whereas the same distance at 100 MHz would not. On the other hand, a distance of 1 mi is insignificant at power-line frequencies but not in the broadcast band.

In this text the various forms of propagation are divided into the following main divisions:

1. *Transmission lines:* two conductors.

2. *Waveguides:* single, hollow conductors.

3. *Antennas:* no conductors in the propagation medium.

---

[1]Richard K. Moore, *Traveling-Wave Engineering* (McGraw-Hill Book Company, 1960).

TABLE 1-1   Free Space-Wavelengths at Various Frequencies

Application	Frequency	Wavelength
Power transmission	60 Hz	3000 mi
Voice	1000 Hz	300 km
Broadcast band	1 MHz	300 m
FM, TV	100 MHz	3 m
X-band radar	10 GHz	3 cm

Each mode has its own application and advantages. Both transmission lines and waveguides have an exponential type of attenuation, with the waveguide having a lower attenuation than a transmission line of equal length. The waveguide however, becomes excessively large at lower frequencies (less than 2 GHz). Antennas have an inverse-square-law type of attenuation characteristic, which is an advantage for longer distances. However, they become very inefficient at low frequencies, where the physical size limitations of an antenna are encountered.

Figure 1-1 shows the relative input power required for the three modes of transmission if a fixed received power is needed. A solid-dielectric coaxial line having a loss of 10 dB/100 ft, a waveguide having a loss of 0.5 dB/100 ft, and

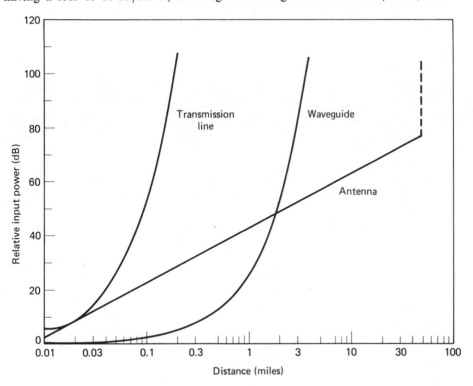

FIG. 1-1   Relative input power required for a fixed receiver power.

transmitting and receiving antennas having a gain of 1000 (30 dB) are assumed at a 2-GHz operating frequency. The antenna has a typical line-of-sight distance. From observation of the figure it is clear that for short transmission distances, waveguides and transmission lines are more efficient, whereas for longer distances, antennas have a clear advantage in efficiency.

1-2 STANDARD TRANSMISSION LINES

Two of the most common wave-guiding structures are the two-wire open line and the coaxial line. These lines generally propagate the "principal mode," which is called the *transverse electromagnetic wave* (TEM). Very simply, this means that the electric (E) and the magnetic (H) fields are always transverse to the direction of wave propagation. If these lines are operated at frequencies that cause the transverse dimensions to become an appreciable portion of a wavelength in size, other, "higher modes" can be set up. These generally are undesirable and are avoided.

Some of the more common types of transmission lines are discussed in this section. The characteristic impedance is also given for each configuration, assuming a lossless transmission line. The significance of this impedance will become evident later. Suffice it to say at this time that it represents the input impedance of a line of infinite length.

Two-Wire Open Line

The *two-wire open line* (Fig. 1-2) is easy to construct and has been extensively used in the telephone industry in the past, and still is very much used in the 60-Hz power industry. Its characteristic impedance (Fig. 1-3) is easy to adjust by changing the spacing. Typical impedances for air-dielectric lines range from about 200 to

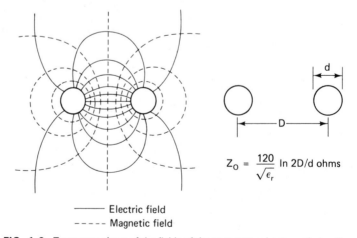

$$Z_0 = \frac{120}{\sqrt{\epsilon_r}} \ln 2D/d \text{ ohms}$$

———— Electric field
----- Magnetic field

FIG. 1-2 Transverse views of the fields of the open two-wire transmission line.

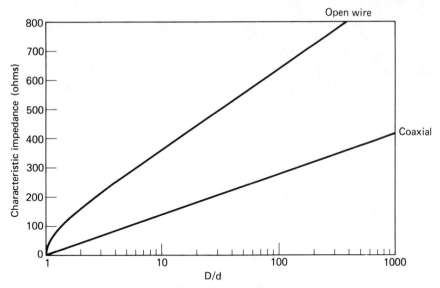

FIG. 1-3 Characteristic impedance of open-wire and coaxial transmission line.

600 Ω. The "twin-lead" transmission lines used commonly in connecting TV receivers to an antenna has a polyethylene supporting structure, resulting in an overall characteristic impedance of 300 Ω.

The two-wire line is a balanced line in that both lines have equal impedances with respect to ground. This differs from the coaxial line, where usually the outer shield or conductor is at ground potential. The real disadvantage of the open-wire line is that its fields extend far beyond the line. This results in excessive radiation losses at the higher frequencies. These lines are seldomly used at frequencies above a few hundred megacycles.

Coaxial Cable

The *coaxial line* (Fig. 1-4) is an unbalanced line (the outer conductor is normally at ground) and probably the most commonly used transmission line. It has no radiation loss but does have a loss depending upon the dielectric used. If air dielectric is used, only slight losses are encountered with the dielectric spacers. The maximum continuous wave power ratings depend upon the size of the conductors and the breakdown voltage of the dielectric in the transmission line.

The *dielectric* commonly used in coaxial lines is polyethylene and more recently polytetrafluorethylene (Teflon). The dielectric constants of some dielectrics are given in Table 1-2. Teflon has superior mechanical strength and melts at higher temperatures but is much more expensive and less pliable.

The *dielectric constant* ϵ is very closely related to a capacitance and for this reason is frequently called *capacitivity*. The units are farads per meter and it has a value in free space of $1/36\pi \times 10^{-9}$ F/m, which is generally symbolized by ϵ_0. The

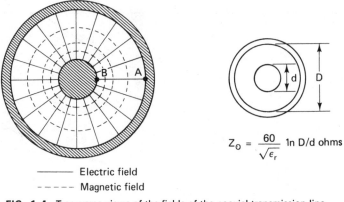

$$Z_0 = \frac{60}{\sqrt{\epsilon_r}} \ln D/d \text{ ohms}$$

——— Electric field

- - - - Magnetic field

FIG. 1-4 Transverse views of the fields of the coaxial transmission line.

TABLE 1-2   Dielectric Constants of Some Dielectrics

Material	Frequency (MHz)	Loss tangent, τ	Relative dielectric constant, ϵ_r	Comments
Polystyrene	100	0.00007	2.6	Plastic. Moderate
	1000	0.0001	2.6	strength. Brittle
	10,000	0.0004	2.5	but machinable. Melts at high temperatures.
Polyethylene	10,000	0.0004	2.3	Plastic. Easily cut, but not machinable; usually molded to shape. Melts at high temperatures.
Teflon	10,000	0.0004	2.1	Tough, machinable, surface has lubricant properties, good high-temperature characteristics.

dielectric constant indicates the amount of charge that can be stored in the dielectric for a given voltage.

The *relative dielectric constant* (ϵ_r) of a material is defined as the ratio of the dielectric constant of the material to the dielectric constant of free space:

$$\epsilon_r = \frac{\epsilon}{\epsilon_0} \tag{1-3}$$

The *power loss* in an insulating material is related to the *conductivity* (σ) of the insulator. To make this a little more meaningful, an insulator can be considered electrically as a large resistance in shunt with a capacitance, as shown in Fig. 1-5. The conductivity (σ) is the inverse of the *resistivity* (ρ).

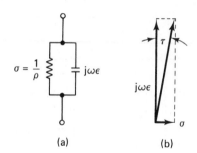

FIG. 1-5 Equivalent circuit of an insulator.

For a good insulator the conductivity will be small. The admittance phase diagram appears in Fig. 1-5(b), where the loss phasor is defined as

$$\tan \tau = \frac{\sigma}{\omega \epsilon}$$

which for small σ reduces to ($\tan \tau \rightarrow \tau$)

$$\tau = \frac{\sigma}{\omega \epsilon} \tag{1-4}$$

The lower the conductivity (the higher the resistivity), the less the attenuation that will be experienced by the signal passing through it.

All lines also experience a resistance loss due to the finite conductivity of the conductors. This loss is generally less than that of the dielectric. All common types of transmission lines experience an increased attenuation with frequency due to skin effect. The attenuation characteristics of one of the better types of coaxial lines are given in Fig. 1-6. This line has the insulating support in the form of a helical ribbon of dielectric material.

These *air-dielectric lines* are often put under slight pressure with nitrogen gas to prevent entry of moisture and other contaminants and to increase the voltage breakdown level. This type of line is used in power transmitter operations, where the power loss must be kept to a minimum. In general, the larger-diameter cables are used for the higher-power applications (see Fig. 1-7), to avoid voltage breakdown under the greater voltages. Great care must be taken when installing or handling these types of lines to avoid crushing, which produces variations in the characteristic impedance of the line. Some lines use a corrugated conductor to improve both flexibility and resistance to crushing, while slightly increasing the attenuation.

For low-power and receiver application, *solid-dielectric lines*, the nominal 50-Ω coaxial cable RG8A/U and RG58A/U, are more common, where the higher attenuation losses are acceptable. Table 1-3 gives the characteristics of some of the coaxial lines that are generally used in communication systems. Solid-dielectric lines are more flexible than air-dielectric lines and also have a higher breakdown voltage.

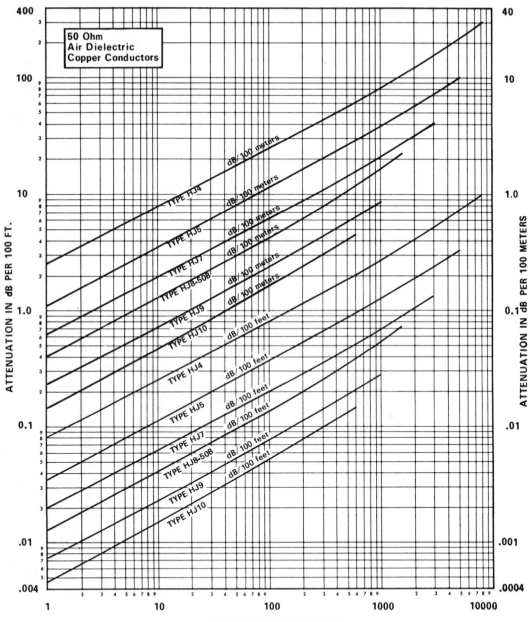

FIG. 1-6 Attenuation of Andrew HELIAX coaxial cable. (From Andrew Catalog 27, containing complete specifications.)

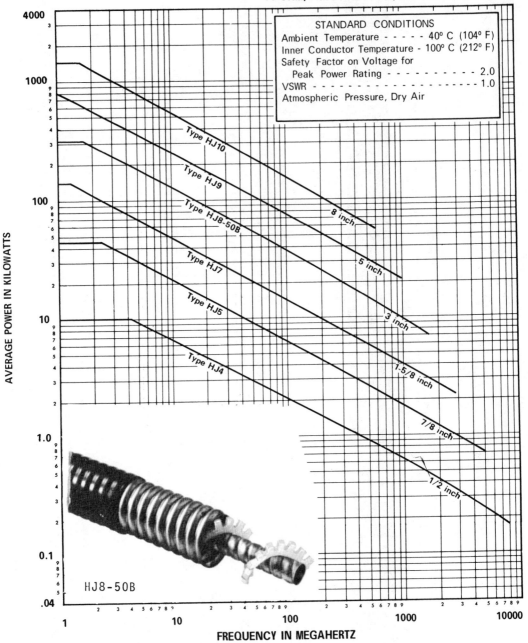

STANDARD CONDITIONS
Ambient Temperature - - - - - 40° C (104° F)
Inner Conductor Temperature - 100° C (212° F)
Safety Factor on Voltage for
    Peak Power Rating - - - - - - - - - - 2.0
VSWR - - - - - - - - - - - - - - - - - 1.0
Atmospheric Pressure, Dry Air

AVERAGE POWER IN KILOWATTS

Type HJ10

Type HJ9

Type HJ8-50B

Type HJ7

Type HJ5

Type HJ4

8 inch

5 inch

3 inch

1-5/8 inch

7/8 inch

1/2 inch

HJ8-50B

FREQUENCY IN MEGAHERTZ

FIG. 1-7 Power rating of Andrew HELIAX coaxial cable. (From Andrew Catalog 27, containing complete specifications.)

TABLE 1-3 Characteristics of Coaxial Lines Used in Communication Systems

COAXIAL CABLES

Description	RG No.	AWG & (Stranding) Material	Insulation Nom. Core O.D. (Inch)	No. of Shields and Material	Jacket	Nom. O.D. (Inch)	Nom. Imp. (Ohms)	Nom. Vel. of Prop.	Nom. Cap. (pF/ft.)	Nom. Attenuation per 100' MHz	db	Standard Spool Lengths in ft.
	8/U JAN-C-17A	13 (7x21) bare copper	Poly-ethylene (.285)	1 bare copper	Black vinyl	.405	52	66%	29.5	100 200 400 900	2.0 3.0 4.7 7.8	50, 100, 500, 1000
	8A/U MIL-C-17D	13 (7x21) bare copper	Poly-ethylene (.285)	1 bare copper	Black non-contaminating vinyl	.405	52	66%	29.5	100 200 400 900	2.0 3.0 4.7 7.8	100, 500, 1000
	8/U Type	11 (7x19) bare copper	Cellular Poly-ethylene (.285)	1 bare copper	Black vinyl	.405	50	78%	26.0	50 100 200 300 400	1.2 1.8 2.6 3.3 3.8	50, 100, 500, 1000
	9/U JAN-C-17A	13 (7x21) silver coated copper	Poly-ethylene (.280)	2-inner-silver coated outer-bare copper	Gray non-contaminating vinyl	.420	51	66%	30.0	100 200 400 900	1.9 2.8 4.1 6.5	100, 1000
	11/U JAN-C-17A	18 (7x26) tinned copper	Poly-ethylene (.285)	1 bare copper	Black vinyl	.405	75	66%	20.5	100 200 400 900	2.0 2.9 4.2 6.5	100, 500, 1000
	11A/U MIL-C-17D	18 (7x26) tinned copper	Poly-ethylene (.285)	1 bare copper	Black non-contaminating vinyl	.405	75	66%	20.5	100 200 400 900	2.0 2.9 4.2 6.5	500, 1000

TABLE 1-3 (continued)

Type	Center conductor	Dielectric	Shield	Jacket	O.D.	Imp.	Vel.	Cap.	Freq. (MHz)	Atten.	Std. lengths
22B/U MIL-C-17D	2 cond. 18 (7x.0152) bare copper, one conductor has tinned center strand	Poly-ethylene (.285)	2 tinned copper	Black non-contaminating vinyl	.420	95	66%	16.0	100 / 200 / 400 / 900	3.0 / 4.5 / 6.8 / 11.0	100, 500, 1000
58/U JAN-C-17A	20 (Solid) bare copper	Poly-ethylene (.116)	1 tinned copper	Black vinyl	.195	53.5	66%	28.5	100 / 200 / 400 / 900	4.1 / 6.2 / 9.5 / 14.5	25, 50, 100, 500, 1000
58A/U JAN-C-17A	20 (19x.0071) tinned copper	Poly-ethylene (.116)	1 tinned copper	Black vinyl	.195	50	66%	30.8	100 / 200 / 400 / 900	5.3 / 8.2 / 12.6 / 20.0	25, 50, 100, 500, 1000
58C/U MIL-C-17D	20 (19x.0071) tinned copper	Poly-ethylene (.116)	1 tinned copper	Black non-contaminating vinyl	.195	50	66%	30.8	100 / 200 / 400 / 900	5.3 / 8.2 / 12.6 / 20.0	500, 1000
58A/U Type	20 (19x32) tinned copper	Cellular Poly-ethylene (.116)	1 tinned copper	Black vinyl	.195	50	78%	26.0	100 / 200 / 400 / 900	4.8 / 6.9 / 10.1 / 15.5	100, 500, 1000
59/U JAN-C-17A	22 (Solid) Copperweld	Poly-ethylene (.146)	1 bare copper	Black vinyl	.242	73	66%	21.0	100 / 200 / 400	3.4 / 4.9 / 7.1	25, 50, 100, 500, 1000
59B/U MIL-C-17D	.023 (Solid) bare Copperweld	Poly-ethylene (.146)	1 bare copper	Black non-contaminating vinyl	.242	75	66%	20.5	100 / 200 / 400 / 900	3.4 / 4.9 / 7.1 / 11.1	500, 1000
59/U Type	22 (6x30) bare (1x29) copper	Cellular poly-ethylene (.146)	1 bare copper	Black vinyl	.242	75	78%	17.3	100 / 200 / 400	2.6 / 3.8 / 5.6	50, 100, 500, 1000

Recommend for camera to recorder to monitor connections. Stranded center conductor adds the flexibility needed to resist severe twisting, bending and other stresses which occur in many CCTV applications.

11

TABLE 1-3 (continued)

Description	RG No.	AWG & (Stranding) Material	Insulation Nom. Core O.D. (Inch)	No. of Shields and Material	Jacket	Nom. O.D. (Inch)	Nom. Imp. (Ohms)	Nom. Vel. of Prop.	Nom. Cap. (pF/ft.)	Nom. Attenuation per 100' MHz	db	Standard Spool Lengths in ft.
	62/U JAN-C-17A	22 (Solid) bare Copperweld	Semi-solid Polyethylene (.146)	1 bare copper	Black vinyl	.242	93	84%	13.5	100 200 400 900	3.1 4.4 6.3 11.0	100, 500, 1000
	62A/U	22 AWG Solid Bare Copper weld	Semi-Solid Polyethylene (.146)	1 bare Copper	Black non-contaminating vinyl	.242	93	84%	13.5	100 200 400 900	3.1 4.4 6.3 11.0	500, 1000, 2000
	62B/U MIL-C-17D	24 (7x32) bare Copperweld	Semi-solid Polyethylene (.146)	1 bare copper	Black non-contaminating vinyl	.242	93	84%	13.5	100 200 400 900	3.1 4.4 6.3 11.0	500, 1000
FR-1	62A/U Type	22 AWG Solid Bare Copperweld	Semi-solid Flame Retardant Polyethylene (.146)	1 bare copper	Black vinyl	.264	93	80%	14.5	100 200 400 900	3.1 4.4 6.3 11.0	500, 1000, 2000
	122/U MIL-C-17D	22 (27x36) tinned copper	Polyethylene (.096)	1 tinned copper	Black non-contaminating vinyl	.160	50	66%	30.8	100 200 400 900	7.0 11.0 16.5 28.0	100, 500, 1000
	141A/U MIL-C-17D	18 (Solid) silver coated Copperweld	TFE TEFLON (.116)	1 silver coated copper	Brown fiber glass	.190	50	69.5%	29.0	400	9.0 Max.	100, 500, 1000
	142B/U MIL-C-17D	18 (Solid) silver coated Copperweld	TFE TEFLON (.116)	2 silver coated copper	Tinted Brown FEP	.195	50	69.5%	29.0	400	9.0 Max.	100, 500, 1000

TABLE 1-3 (continued)

	Cable	Conductor	Dielectric	Shield	Jacket	Nom. O.D.	Imped.	Vel.	Cap.	Freq. (MHz)	Atten.	Length (ft)
	174/U MIL-C-17D	26 (7x34) bare Copperweld	Poly-ethylene (.060)	1 tinned copper	Black vinyl	.100	50	66%	30.8	100 200 400	8.8 13.0 20.0	100, 500, 1000
	178B/U MIL-C-17D	30 (7x38) silver coated Copperweld	TFE TEFLON (.034)	1 silver coated copper	Tinted Brown FEP	.070	50	69.5%	29.0	400	29.0 Max.	100, 500
	179B/U MIL-C-17D	30 (7x38) silver coated Copperweld	TFE TEFLON (.063)	1 silver coated copper	Tinted Brown FEP	.100	75	69.5%	19.5	400	21.0 Max.	100, 500, 1000
	180B/U MIL-C-17D	30 (7x38) silver coated Copperweld	TFE TEFLON (.102)	1 silver coated copper	Tinted Brown FEP	.140	95	69.5%	15.0	400	17.0 Max.	100, 500
	213/U MIL-C-17D	13 (7x21) bare copper	Poly-ethylene (.285)	1 bare copper	Black non-contaminating vinyl	.405	50	66%	30.8	100 200 400 900	2.0 3.0 4.7 7.8	500, 1000
	214/U MIL-C-17D	13 (7x.0296) silver coated copper	Poly-ethylene (.285)	2 silver coated copper	Black non-contaminating vinyl	.425	50	66%	30.8	100 200 400 900	2.0 3.0 4.7 7.8	500, 1000
	223/U MIL-C-17D	19 (Solid) silver coated copper	Poly-ethylene (.116)	2 silver coated copper	Black non-contaminating vinyl	.206	50	66%	30.8	100 200 400 900	4.8 7.0 10.0 16.0	100, 500, 1000
	316/U MIL-C-17D	26 (7x.0067) silver coated Copperweld	TFE TEFLON (.060)	1 silver coated copper	Tinted Brown FEP	.098	50	69.5%	29.0	400	20.0 max.	100, 500, 1000

Source: From Belden Electronic Wire and Cable Catalog 873 containing complete specifications.

Great care should always be taken when attaching connectors to lines to avoid introducing discontinuity resulting in mismatching or a possible voltage breakdown in high-voltage applications.

Balanced Shielded Line

To obtain a *balanced line* and also to maintain low radiation losses, a shielded pair of wires can be employed (Fig. 1-8). This necessarily increases the costs of the cable.

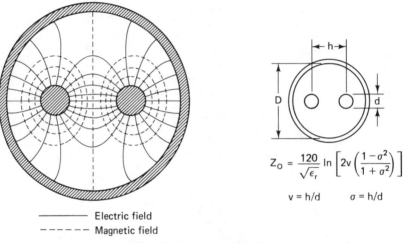

——————— Electric field

－ － － － － Magnetic field

FIG. 1-8   Balanced shielded line.

Tri-Plate Line

The *tri-plate line* shown in Fig. 1-9 consists of a strip conductor bounded by a dielectric material, which, in turn, is confined by two ground planes. The field intensity falls off rapidly with distance from the conducting strip. These lines can be manufactured using standard photo-etching techniques and are used to interconnect components that operate up to frequencies of 10 GHz.

Microstrip Line

The *microstrip transmission line* illustrated in Fig. 1-10(a) consists of a conductor above a single ground plane. By employing a high dielectric substrate, the electromagnetic field is concentrated very tightly near the conductor in the free-space region. Because some of the field lines are outside the substrate, the effective dielectric constant is somewhat lower than that of the substrate. An alumina dielectric, which is a ceramic material having a relative dielectric constant of approximately 10, is often employed. The result is a very compact circuit which can be

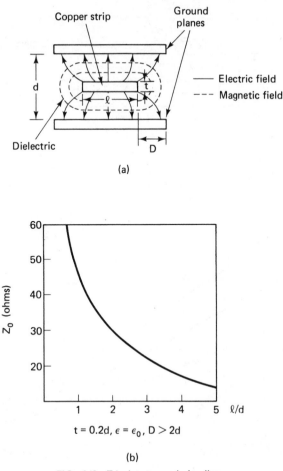

(b)

FIG. 1-9   Tri-plate transmission line.

produced at very reasonable costs. It also provides very convenient mounting of discrete components on the microstrip line.

The high-dielectric-constant substrates permit nearby circuit traces to be positioned very closely without danger of interference or radiation. For characteristics impedance of up to about 50 Ω, neighboring traces should be spaced a minimum of one microstrip width away. This should be increased for higher impedances.

Parallel Line

The *parallel line* shown in Fig. 1-11 is frequently used in situations where low impedances are required, as in high-current applications.

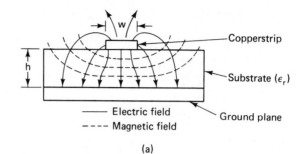

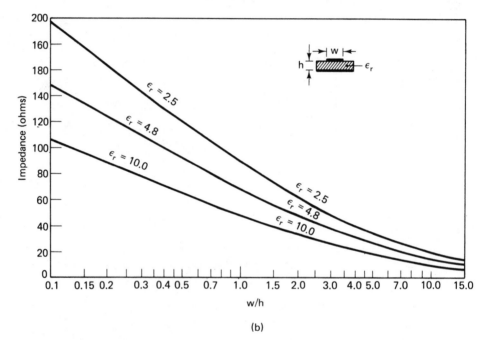

(b)

FIG. 1-10 Microstrip transmission line; (a) microstrip line; (b) characteristic impedance of a microstrip transmission line for glass epoxy (ϵ_r = 4.8), Teflon epoxy (ϵ_r = 2.55), and alumina (ϵ_r = 10). For line widths much greater than dielectric thickness ($w \gg h$) the characteristic impedance can be approximated by $Z_0 = (377/\sqrt{\epsilon_r})\,(h/w)$. (From James R. Fisk, "Microstrip Transmission Line," *Ham Radio Magazine*, January 1978, Fig. 5.)

1-3 VELOCITY OF PROPAGATION IN TRANSMISSION LINES

The velocity of wave propagation in a transmission line having an air dielectric is equal to the velocity of light, c. If, however, a dielectric material such as polyethylene is employed, as in RG8 and RG58, then the velocity of propagation is

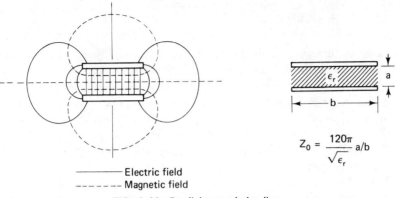

————— Electric field
– – – – – Magnetic field

FIG. 1-11 Parallel transmission line.

$$Z_0 = \frac{120\pi}{\sqrt{\epsilon_r}}\,a/b$$

less than that of light. The velocity then is equal to (assuming a lossless line)

$$v = \frac{c}{\sqrt{\mu_r \epsilon_r}} \qquad (1\text{-}5)$$

where ϵ_r is the relative dielectric constant and μ_r the relative permeability constant. The relative permeability constant is the ratio of the permeability of the material under consideration (μ) to the permeability of free space ($\mu_0 = 4\pi \times 10^{-7}$ H/m). As one can observe from the units, the permeability constant appears to be an inductive type of parameter and for that reason is at times called *inductivity*.

For most materials (not including the ferrous compounds) μ_r is taken to be 1. This is particularly true of the plastics, used as dielectrics in transmission lines. Under these circumstances,

$$v = \frac{c}{\sqrt{\epsilon_r}}$$

Let us now consider the problem of obtaining the velocity and wavelength of a 100-MHz signal in RG8 cable, and in free space. This is done in tabular form in Table 1-4. The velocity of propagation of a wave in RG8/U is seen to be 66% ($1.98/3 \times 100$) of that of the velocity of light. This is seen to be in agreement with the manufacturer's data in Table 1-3.

TABLE 1-4 Wavelength and Velocity in RG8 at 100 MHz

	Free space	*RG8* ($\epsilon_r = 2.3$)
v	3×10^8 m/s	$\dfrac{c}{\sqrt{\epsilon_r}} = 1.98 \times 10^8$ m/s
λ	3 m	$\dfrac{v}{f} = 1.98$ m

1-4 SELECTION OF CHARACTERISTIC IMPEDANCE OF A COAXIAL TRANSMISSION LINE

For a coaxial line with small losses, the characteristic impedance depends only upon the diameters of the conductors and the dielectric constant of the medium between the concentric cylinders. The selection of a characteristic impedance depends upon the parameter one wishes to optimize. For an air-dielectric coaxial line, for instance, the maximum power-carrying capacity occurs at a diameter ratio (D/d of Fig. 1-4) of 1.65, which results in a characteristic impedance of 30 Ω. On the other hand, if the voltage breakdown is optimized, a diameter ratio of 2.7 is obtained, which corresponds to a characteristic impedance of 60 Ω.

Both of the cases above give the current density in the conductors, which figures strongly when considering attenuation losses. Minimum attenuation is achieved when the diameter ratio reaches 3.6 or a characteristic impedance of 77 Ω. Figure 1-12 graphically depicts how the power-carrying capacity, breakdown voltage, and attenuation varies with the coaxial-line characteristic impedance. The 77 Ω impedance has been standardized to 75 Ω and is in very much use as, for example, in RG11/U cable and the UHF connector. If one fills the air space in the

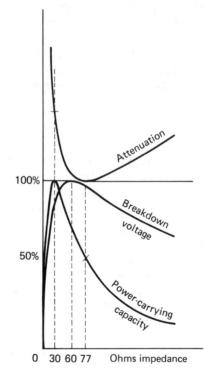

FIG. 1-12 Electrical parameter variations with the characteristic impedance of an air-dielectric coaxial line.

$77\,\Omega$ air line with a low-loss dielectric such as polyethylene, the characteristic impedance is reduced to

$$Z_0 = \frac{77}{\sqrt{\epsilon_r}} = 51\,\Omega$$

This impedance, along with $50\,\Omega$, is also in common use. RG8/U and RG58/U are examples. The GR line with GR connectors are also $50\,\Omega$ lines. The type N connector is a $50\,\Omega$ connector.

PROBLEMS

1-1. A transmission line with air dielectric is 25 m long. How long is the line in wavelengths at frequencies of:
   (a) 1000 Hz?
   (b) 10 MHz?
   (c) 100 MHz?

1-2. A transmission line with a dielectric ($\epsilon_r = 3.5$) is 100 m long. How many wavelengths long is the line at a frequency of 10 GHz?

1-3. (a) Why are waveguides not used at low frequencies?
   (b) Why are open-wire lines not generally used as guiding structures at very high frequencies?
   (c) What is the velocity of wave propagation in a Teflon coaxial transmission line?

1-4. Plot on log-log paper the attenuation characteristics of the following types of coaxial lines:
   (a) RG8A/U.
   (b) RG58A/U.
   (c) Andrew HJ7 rigid line.

1-5. It is desired to cut a $\lambda/4$ length of RG58A/U cable at 150 MHz. What is the physical length of this cable?

1-6. What is the conductivity of Teflon at 10 GHz?

two

TRANSIENTS
ON A LOSSLESS
TRANSMISSION LINE

2-1 DISTRIBUTED CONSTANTS
OF A LOSSLESS TRANSMISSION LINE

In order to determine the voltage and current along a transmission line, we must establish the electrical characteristics or the electrical equivalent of the line. The two most important parameters of the transmission line (the only parameters considered in a lossless line) are the inductance and capacitance. The current in the line sets up a magnetic flux around the conductors, which in turn induces a voltage in the conductors $\left(\text{the familiar } L\frac{di}{dt}\right)$. This distributed *inductance* is represented by the symbol L, having units of henries per unit length of line. The *capacitance*, which is also distributed (think of the lines as two parallel plates) is represented by C. This is measured in farads per unit length of line.

A simplistic schematic of the distributed L and C is given in Fig. 2-1. As a current–voltage wave (or an electromagnetic wave) travels down a transmission line, the currents must flow through the distributed inductors and a voltage must be set up across the distributed capacitance. Because the current cannot

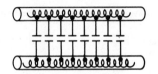

FIG. 2-1   Equivalent circuit of a lossless transmission line.

change instantaneously through an inductor, and a voltage cannot be immediately changed across a capacitor, it takes time for the current–voltage wave to travel down the line. Thus, there is a finite velocity for the wave propagation down the line.

   If neither the series resistance of the line nor the finite conductance of the insulation material cannot be neglected, there is also an attenuation of the wave as it progresses down the lossy line. We shall neglect these parameters for the moment but will consider them later.

2-2   TRAVELING WAVES ON A LOSSLESS TRANSMISSION LINE

It is possible to mathematically derive the general solutions for the voltage and current along a uniform transmission line. We shall do this later for the sinusoidal steady-state case, but since the general solution involves solving a partial second-order differential equation, we shall for the moment merely describe the traveling-wave phenomenon of a transient voltage or current on a transmission line.

   The convention used when dealing with voltages and currents on transmission lines is slightly different from that used in two-port circuit theory. The voltage is considered to be positive when the upper terminal or lead is at a more positive potential than the lower lead, and the currents are taken to be positive when traveling from the generator to the load. This is illustrated in Fig. 2-2, where the subscripts S and R refer to the sending end and the receiving end, respectively.

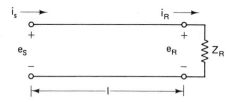

FIG. 2-2   Convention used in transmission-line theory.

   Often, when a source is applied to such a line, two distinct waves are present, which are called the *incident wave* and the *reflected wave*. The incident wave propagates from the source to the receiving end, whereas the reflected wave propagates from the receiving end toward the sending end. This is illustrated in Fig. 2-3, where, by convention, a current traveling toward the generator is taken to be negative. Rigorous analysis, as well as experimental data, indicates that these traveling

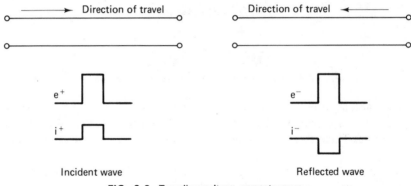

FIG. 2-3   Traveling voltage–current waves.

waves, whether incident or reflected, see what is called the *characteristic impedance* (Z_0) of the transmission line. The characteristic impedance of several types of transmission lines were given in Section 1-2. This impedance is independent of the line length or load and is a function of the line parameters only (i.e., size and spacing of conductors and type of insulation used).

We shall denote the incident voltage wave by e^+, which is a wave that travels from the source to the load. Similarly, the reflected voltage wave will be denoted by e^-. The incident and reflected current waves will be symbolized by i^+ and i^-, respectively.

Since a traveling wave sees the characteristic impedance of the transmission line, the incident current is related to the incident voltage by the relation

$$i^+ = \frac{e^+}{Z_0} \tag{2-1}$$

The reflected current can be expressed in terms of the reflected voltage as

$$i^- = -\frac{e^-}{Z_0} \tag{2-2}$$

where the negative sign appears due to the convention of positive current traveling toward the load (this current flows in the direction opposite to the incident current). The question that probably comes to mind now is: What causes a reflected wave to exist on a line? It is not too difficult to visualize that when a generator introduces a signal on the line, a wave will be set up which will travel away from the source; but what causes a wave to be present that travels in the reverse direction? To see this a little more clearly, let us consider a wave traveling toward the load. This wave, as it moves along the line, sees the characteristic impedance of the line. If the line is terminated with an impedance having a value equal to the characteristic impedance of the line ($Z_R = Z_0$), the wave will not notice any change as it reaches the termination, and the total voltage can be taken to be the incident voltage (the same holds for the current). In other words, no reflections need to occur, as Ohm's law is still validated. That is, when $Z_R = Z_0$,

$$\frac{e_R^+}{i_R^+} = Z_R = Z_0$$

The subscript R is used here to denote currents, voltages, and so on, at the receiving end. *If, however, the load impedance is not equal to the characteristic impedance of the line, another wave must be set up to assure that Ohm's law is obeyed.*

   The voltage at the load divided by the current through it must equal the load impedance. Since

$$\frac{e_R^+}{i_R^+} = Z_0 \neq Z_R$$

in this case, a reflection must occur.

   At the termination we must have

$$\frac{\text{total } e}{\text{total } i} = Z_R \tag{2-3}$$

Unless Z_R is equal to Z_0, the incident wave alone does not satisfy this relation and a reflected wave must occur. The total voltage and current at the load can be expressed as

$$\text{total } e = e_R^+ + e_R^- \tag{2-4}$$

$$\text{total } i = i_R^+ + i_R^- = \frac{e_R^+}{Z_0} - \frac{e_R^-}{Z_0} \tag{2-5}$$

Substituting equation (2-5) into equation (2-3), we obtain

$$\frac{\text{total } e}{\text{total } i} = Z_R = Z_0 \frac{e_R^+ + e_R^-}{e_R^+ - e_R^-} \tag{2-6}$$

From this equation, the relation between the reflected voltage and the incident voltage can be found:

$$\frac{e_R^-}{e_R^+} = \frac{Z_R - Z_0}{Z_R + Z_0} = \Gamma_R \tag{2-7}$$

The ratio Γ_R is called the *voltage reflection coefficient*. If $Z_R = Z_0$, it is noted that Γ_R goes to zero, and no reflected wave is present:

$$\Gamma_R = \frac{Z_R - Z_0}{Z_R + Z_0} = \frac{Z_0 - Z_0}{Z_0 + Z_0} = 0$$

If Γ_R is not equal to zero, a reflected wave does occur, having a value equal to

$$e_R^- = e_R^+ \Gamma_R$$

Similarly, it can be proven that the current reflection coefficient (i_R^-/i_R^+) is the negative of that for the voltage reflection coefficient. Some examples will now be considered to illustrate the concepts just outlined. *All the transmission lines are assumed to be lossless.*

EXAMPLE 2-1

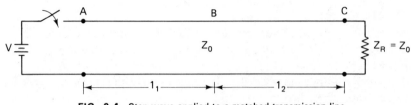

FIG. 2-4 Step wave applied to a matched transmission line.

A battery, having a voltage of magnitude V, is applied to a transmission line at time $t = 0$. Determine the waveforms that will be present at locations A, B, and C. Assume the velocity of propagation to be the velocity of light, c.

Solution: When the switch closes, a voltage wave will commence to move down the line at a velocity c. It arrives at point B in l_1/c seconds and at point C in $(l_1 + l_2)/c$ seconds. No reflection is present as the line is properly terminated (i.e., $Z_R = Z_0$).

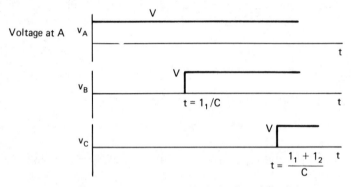

FIG. 2-5 Waveforms on the transmission line of Fig. 2-4.

EXAMPLE 2-2

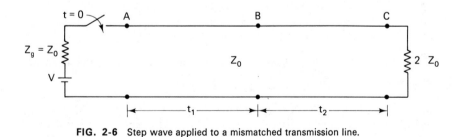

FIG. 2-6 Step wave applied to a mismatched transmission line.

A battery, having a voltage V and an internal impedance of Z_0, is applied to a transmission line at time $t = 0$. Determine the waveforms that will be present at locations A, B, and C.

Assume t_1 to be the time taken for the wave to arrive at point B and t_2 to be the time taken to travel the distance from B to C.

Solution: When the switch closes, a voltage wave commences moving down the transmission line, seeing an impedance Z_0. Since there will be a voltage drop across the generator impedance, the magnitude of the voltage moving down the line is not equal to the battery voltage. The equivalent circuit for the wave as it just enters the line consists of a generator with its internal impedance Z_0 and the characteristic impedance of the line Z_0.

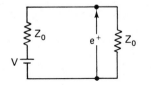

FIG. 2-7 Equivalent input circuit at $t = 0$.

The incident voltage has a magnitude of

$$e^+ = V \frac{Z_0}{Z_0 + Z_0} = \frac{V}{2}$$

The reflection coefficient at the receiving end has a value of

$$\Gamma_R = \frac{Z_R - Z_0}{Z_R + Z_0} = \frac{2Z_0 - Z_0}{2Z_0 + Z_0} = \frac{1}{3}$$

The reflected voltage has a magnitude of

$$e^- = \Gamma_R e^+ = \frac{1}{3} \times \frac{V}{2} = \frac{V}{6}$$

The waveforms are shown in Fig. 2-8. The receiving end is matched, and there will be no further reflections.

EXAMPLE 2-3

A battery is applied to a transmission line at time $t = 0$, as indicated in Fig. 2-9. Determine the waveforms at A, B, and C. Assume t_1 and t_2 to be the times it would take a wave to travel from point A to B and from point B to C, respectively.

Solution: As explained in Example 2-2, a voltage of magnitude $V/2$ commences to move down the line upon closing the switch. When this wave reaches point B, it experiences a mismatch, since it effectively sees the shunting resistor (R) in parallel with the Z_0 of the continuing section of transmission line. The impedance at point B as seen by the wave is $Z_0/2$; and the reflection coefficient at this point, noted by Γ_B, is equal to

$$\Gamma_B = \frac{Z_0/2 - Z_0}{Z_0/2 + Z_0} = -\frac{1}{3}$$

The reflected wave from this point will then have a value of

$$e_B^- = \Gamma_B e_B^+ = -\frac{1}{3} \times \frac{V}{2} = -\frac{V}{6}$$

This wave will be reflected toward the generator.

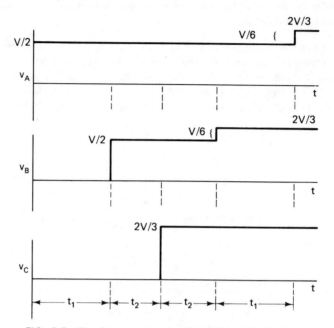

FIG. 2-8 Waveforms on the transmission line of Fig. 2-7.

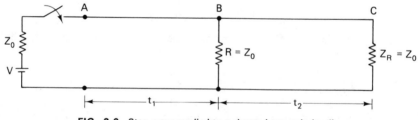

FIG. 2-9 Step wave applied to a shunted transmission line.

The total voltage remaining at point **B**, which will also begin moving toward the load is

$$e_B^+ + e_B^- = \frac{V}{2} - \frac{V}{6} = \frac{V}{3}$$

Both the receiving and generator ends are matched and there will be no further reflections. Figure 2-10 shows the resulting waveforms.

The final steady-state voltage or dc condition can be checked by assuming the transmission line to be merely a pair of connecting leads (Fig. 2-11). The dc voltage checks out to be

$$V_{dc} = V_A = V_B = V_C = V \times \frac{Z_0/2}{Z_0 + Z_0/2} = \frac{V}{3} \quad \text{volts}$$

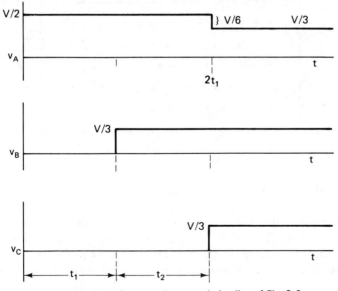

FIG. 2-10 Waveforms on the transmission line of Fig. 2-9.

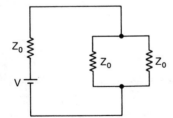

FIG. 2-11 Steady-state voltage on the transmission line of Fig. 2-9.

In either Example 2-2 or 2-3, if the generator impedance is also mismatched to the line, multiple reflections would occur. The reflected voltage at the generator end would have a value of

$$e_s^- = \Gamma_s e_s^+$$

where e_s^+ is the voltage reflected back from the receiving end and Γ_s is the reflection coefficient the generator would present to the line:

$$\Gamma_s = \frac{Z_g - Z_0}{Z_g + Z_0}$$

Figure 2-12 indicates the sequence of reflections that would occur on such a line, where the second subscript, after the parentheses, represents the number of reflections that have occurred at the respective end.

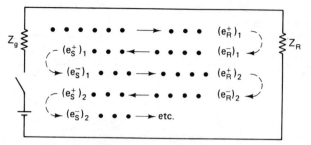

FIG. 2-12 Reflections on a totally mismatched line.

2-3 REFLECTIONS FROM REACTIVE LOADS

Let us now consider the case where a step wave is applied to a lossless line terminated in an inductor (Fig 2-13). When the leading edge of the step wave arrives at the load, the inductor appears as an open circuit, as the frequencies present in this portion of the wave are extremely high ($X_L = 2\pi f L$). As time progresses,

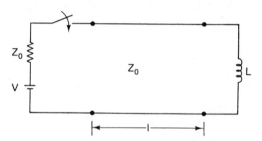

FIG. 2-13 Transmission line with an inductive load.

however, the step wave settles down to a dc voltage and the inductor then appears as a short. An oscilloscope sees a waveform at the sending end as that sketched in Fig. 2-14. The delay in Fig. 2-14 is due to the travel time of the wave up and down the transmission line.

If a capacitor forms the load, the opposite will occur. The high frequencies present at the leading edge of the wavefront see a very low reactance ($X_C = 1/2\pi f C$), whereas the later dc condition sees a very high impedance. Figure 2-15 shows the waveform that will be observed at the input of such a terminated line.

2-4 TIME-DOMAIN REFLECTOMETRY

Many test instruments are now being manufactured which employ a step voltage or a bell-shaped pulse to detect the location of a discontinuity in a transmission system. The Hewlett-Packard Model 1415A *time-domain reflectometer* (TDR), for

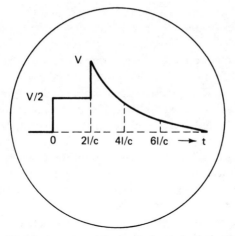

FIG. 2-14 Waveform as seen at the input of an inductively terminated line.

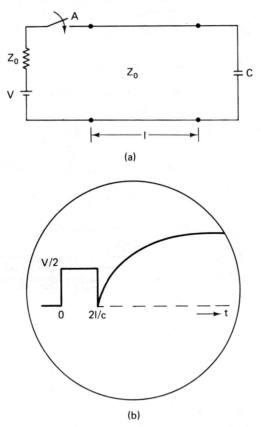

(a)

(b)

FIG. 2-15 (a) Transmission line with a capacitive load; (b) Waveform as seen at point A of Fig. 2-15(a).

instance, employs a 150-ps rise-time step voltage and can locate faults within a distance of a few centimeters. Simply stated, this TDR applies a voltage step to the transmission system and the reflections are observed. A reflection occurs each time the step encounters a discontinuity; this reflection is added to the incident wave and is displayed on the cathode-ray-tube oscilloscope. The time required for the reflection to return to the oscilloscope locates the discontinuity. The shape and magnitude of the reflected wave indicate the nature and value of the mismatch, which can be resistive, inductive, or capacitive. In general, an inductive discontinuity reflects a voltage spike having the same polarity as the incident step, and a capacitive discontinuity reflects a voltage spike of the opposite polarity. A resistive discontinuity having a value larger than the line impedance reflects a step of the same polarity as the incident step, and a step of the opposite polarity is reflected if the value is less than the line impedance.

When long lossy cables are tested with a time-domain reflectometer, both the amplitude and the shape of the reflections are changed. In general, the rise time is badly degraded and distance measurements are somewhat obscured. The error can be greatly reduced if the distance between the 10% points on the leading edge of two reflections are used rather than the distance between the peaks of the reflections. This is shown in Fig. 2-16.

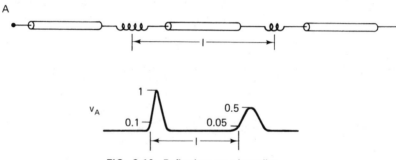

FIG. 2-16   Reflections on a lossy line.

Time-domain reflectometry can only be usefully employed in *broad-band systems* also having a dc response. Single-conductor waveguides, for instance, do not transmit frequencies below a *cutoff frequency* and therefore cannot be analyzed by this technique.

PROBLEMS

2-1. Define the voltage reflection coefficient.

2-2. Originally the voltage is zero over the entire length of the transmission line. At $t = 0$ the switch closes. Find the voltage waveforms at A and B.

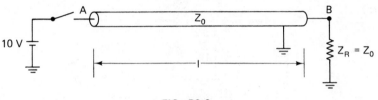

FIG. P2-2

2-3. (a) Originally, the voltage is zero over the entire length of the transmission line. At time $t = 0$ the switch closes. Find the voltage waveforms at A and B.

(b) If the load Z_R is removed such that an open-circuit condition results at point B, what would the new voltage waveforms be at A and B?

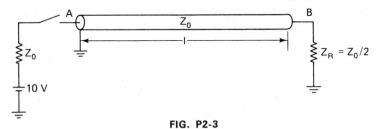

FIG. P2-3

2-4. The following voltage waveform is seen on an oscilloscope connected to the input of a length of faulted RG8A/U line. Determine the fault location (from the sending end) and the fault impedance. Ignore any losses.

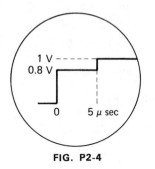

FIG. P2-4

2-5. The following waveforms are observed at an oscilloscope connected to the input of a cable loaded with an unknown impedance. Circle the correct answer.

(a)

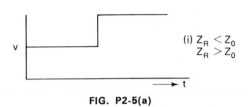

FIG. P2-5(a)

(b)

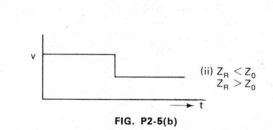

(ii) $Z_R < Z_0$
$Z_R > Z_0$

FIG. P2-5(b)

(c) What is the generator impedance relative to Z_0 of the cable for the above cases.

2-6. The following voltage waveform is seen on a time-domain reflectometer ($R_G = 50\ \Omega$) connected to a shorted section of coaxial cable.
(a) If the dielectric material is Teflon ($\epsilon_r = 2.1$), determine the length of cable.
(b) Calculate the characteristic impedance of the cable. (Assume no losses.)

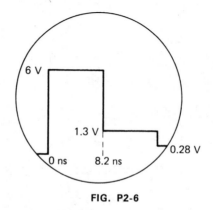

FIG. P2-6

2-7. The following voltage waveform is seen on a time-domain reflectometer ($R_G = 50\ \Omega$) connected to an open section of coaxial cable. If the dielectric material is polystyrene ($\epsilon_r = 2.5$), determine the:
(a) Length of the cable.
(b) Characteristic impedance of the cable. (Assume no losses.)

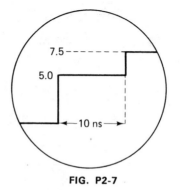

FIG. P2-7

three

STEADY-STATE
CONDITIONS
ON A
TRANSMISSION LINE

3-1 ROTATING PHASOR

Sinusoidal oscillations are generally expressed in the form

$$e = E_m \cos(\omega t + \phi) \tag{3-1}$$

where t = time
   ω = angular frequency $(2\pi f)$
   ϕ = phase angle (radians)
   E_m = maximum amplitude
   e = instantaneous voltage

This equation can be graphically sketched by noting that its maximum amplitude is equal to E_m and the phase shift (also called displacement) is equal to $\omega t = -\phi$.

The period of this expression in ωt is equal to 2π. The waveform is shown in Fig. 3-1. When dealing with traveling waves, this is not the most convenient mathematical representation of a sinusoidal signal. For this reason, the rotating phasor notation is usually employed (see Fig. 3-2).

When representing voltage and currents on transmission

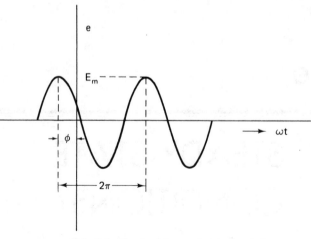

FIG. 3-1   Graph of $E_m \cos{(\omega t + \phi)}$.

lines (E and H fields as well), the rotating phasor is expressed as (see Appendix A)

$$E_m e^{j(\omega t + \phi)}$$

where

$$e^{j(\omega t + \phi)} = \cos{(\omega t + \phi)} + j \sin{(\omega t + \phi)} \qquad (3\text{-}2)$$

$E_m = peak$ magnitude of this expression

$e =$ base of the natural logarithm and ϕ represents the phase angle

If we project the rotating phasor onto the horizontal line (real axis), it is evident that this projection is given by

$$e = E_m \cos{(\omega t + \phi)}$$

which is identical to expression (3-1). Therefore, the instantaneous voltage equation (or current) can be expressed in terms of a rotating phasor, such as

$$e = \text{Re}\,[E_m e^{j(\omega t + \phi)}] \qquad (3\text{-}3)$$

where the symbol Re is used as "the real part of." This notation is frequently used because of its convenient form (which will be seen more clearly later) and because of the fact that the exponential does not change its mathematical form when differentiated or integrated.

Equation (3-3) can be rewritten in the form

$$e = \text{Re}\,[(E_m e^{j\phi})e^{j\omega t}] \qquad (3\text{-}4)$$

The quantity in parentheses, $E_m e^{j\phi}$, is a complex number which represents the *peak phasor* of the voltage e.

If a sinusoidal voltage (or current) is given as a root-mean-square value, it can be converted to a peak value by multiplying by $\sqrt{2}$:

$$E_m = \sqrt{2}\,E_{\text{rms}}$$

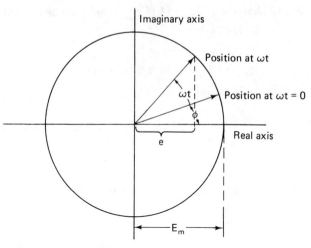

FIG. 3-2 Rotating phasor.

It is common practice to omit the symbol Re and to write simply

$$e = E_m e^{j\phi} e^{j\omega t} \tag{3-5}$$

When this is done, one must not forget that the instantaneous voltage e is, in fact, the projection of the rotating phasor on the real axis.

A few examples will now be considered.

EXAMPLE 3-1

Obtain the peak phasor of the quantity

$$e = 10 \cos (\omega t + \pi/8)$$

Solution: This expression can be written in the general form of equation (3-4):

$$e = \text{Re} \underbrace{(E_m e^{j\phi} e^{j\omega t})}_{\substack{\text{peak} \\ \text{phasor}}}$$

$$= \text{Re} \left[10 e^{j(\omega t + \pi/8)}\right]$$

$$= \text{Re} \left(10 e^{j\pi/8} e^{j\omega t}\right)$$

The peak phasor is $10 e^{j\pi/8} = 10 \,\underline{/\pi/8}$.

EXAMPLE 3-2

Obtain the instantaneous value of current for the rms phasor.

$$I = 10\underline{/30°} \text{ A at } \omega t = 20° \qquad (30° = \pi/6 \text{ rad})$$

Solution: The peak phasor is $\sqrt{2}I = 14.14\underline{/30°}$ A. Substituting into equation (3-4),

$$i = \text{Re}\,(14.14e^{j\pi/6}e^{j\omega t})$$
$$= \text{Re}\,[14.14e^{\,j(\omega t + \pi/6)}]$$
$$= \text{Re}\,\{14.14[\cos(\omega t + \pi/6) + j\sin(\omega t + \pi/6)]\}$$
$$= 14.14\cos(\omega t + \pi/6)$$

At $\omega t = 20°$,

$$i = 14.14\cos(20° + 30°) = 14.14\cos 50°$$
$$= 9.1 \text{ A}$$

3-2 DERIVATION OF THE DIFFERENTIAL STEADY-STATE EQUATIONS FOR THE UNIFORM TRANSMISSION LINE

In Fig. 2-1 the equivalent circuit of a uniform lossless transmission line is given where only the distributed L and C are considered. In an actual line the resistance of the conductors and the conductance of the insulation must also be taken into account. The resistance of the conductors (which can also account for radiation losses in an open-wire line) is denoted by the resistance R, measured in ohms per unit length of line. The imperfection of the insulator (dielectric loss) is denoted by the symbol G, having units of Siemens per unit length of line. One should note that G is not equal to the reciprocal of R, since one refers to the dielectric loss, whereas the other refers to the copper loss.

The notation and units that will be used are visualized in Fig. 3-3. The subscripts S and R refer to the sending end and receiving end, respectively; x is the distance from the sending-end terminals and d represents the distance from the receiving or load end. Again, by convention, currents are taken to be positive when flowing toward the load.

Let us now consider a short section (Δx) of transmission line at the location x. We will represent this short section of line by a model where all the series elements are combined on one side ($R\,\Delta x$ and $L\,\Delta x$), and where the shunt elements

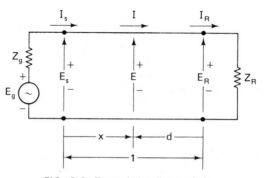

FIG. 3-3 Transmission-line symbols.

($G \Delta x$ and $C \Delta x$) are shown on the right-hand side of the section. Other models are also used (e.g., T and π models), but each one results in the same transmission-line equations. Our model is shown in Fig. 3-4, where the four parameters of the line are

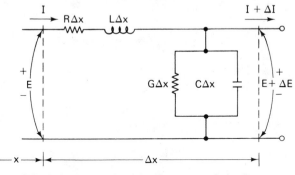

FIG. 3-4 Elemental length of lossy transmission line.

$R:$ series resistance per unit length.
$L:$ series inductance per unit length.
$G:$ shunt conductance per unit length.
$C:$ shunt capacitance per unit length.

A line of length Δx has a series resistance of $R \Delta x$ ohms and series inductance of $L \Delta x$ henrys. Similarly, the shunt conductance is $G \Delta x$ Siemens and the shunt capacitance is $C \Delta x$ farads.

By employing Ohm's and Kirchhoff's voltage law, we can relate the output voltage to the input voltage by the expression

$$E + \Delta E = E - I(R + j\omega L) \Delta x$$

or

$$\frac{\Delta E}{\Delta x} = -(R + j\omega L)I \tag{3-6}$$

This equation states that the voltage change (ΔE) that occurs in a distance (Δx) is due to the voltage drop in the series impedance ($R + j\omega L$).

Similarly, the difference in current between the two ends of the section is due to the shunting of the current through $G \Delta x$ and $C \Delta x$. Thus,

$$\Delta I + I = I - (G + j\omega C) \Delta x(E + \Delta E)$$
$$= I - (G + j\omega C) \Delta x E$$

where the higher-order term $\Delta x \, \Delta E$ can be ignored for small Δx (then ΔE is also small).

This equation can be rewritten as

$$\frac{\Delta I}{\Delta x} = -(G + j\omega C)E \tag{3-7}$$

Equation (3-7) indicates that the current change (ΔI) along a length of transmission line (Δx) is due to the shunting effect of $G + j\omega C$.

The differential equations for the voltage–current on a transmission line can now be derived by letting Δx approach zero in equations (3-6) and (3-7).

$$\lim_{\Delta x \to 0} \frac{\Delta E}{\Delta x} = \frac{dE}{dx} = -(R + j\omega L)I \tag{3-8}$$

$$\lim_{\Delta x \to 0} \frac{\Delta I}{\Delta x} = \frac{dI}{dx} = -(G + j\omega C)E \tag{3-9}$$

It is usual to denote the ac series impedance per unit length ($R + j\omega L$) by the symbol Z and the shunt admittance per unit length by the symbol Y.

$$Z = R + j\omega L \quad \text{ohms/unit length} \tag{3-10}$$

$$Y = G + j\omega C \quad \text{Siemens/unit length} \tag{3-11}$$

The differential equations can then be written more compactly as

$$\frac{dE}{dx} = -ZI \tag{3-12}$$

$$\frac{dI}{dx} = -YE \tag{3-13}$$

ac Steady-State Solution for the Uniform Line

To obtain the expressions for the voltage and current along a transmission line, we must solve the previously derived differential equations (3-12) and (3-13). We will first solve for the voltage E by eliminating I from equation (3-12). This can be accomplished by differentiating equation (3-12) with respect to x and substituting for the resultant dI/dx, equation (3-13).

The derivative of equation (3-12) is

$$\frac{d^2E}{dx^2} = -Z\frac{dI}{dx}$$

Substituting equation (3-13) into this for dI/dx we obtain

$$\frac{d^2E}{dx^2} = (YZ)E \tag{3-14}$$

The solution of this equation can take many forms (hyperbolic functions, complex sinusoidal functions, exponential functions, etc.), from which we pick the convenient exponential form. The general solution of equation (3-14) is expressed as

$$E = A_1 e^{-\sqrt{ZY}x} + A_2 e^{\sqrt{ZY}x} \tag{3-15}$$

where the A's are the constants of integration (which must still be evaluated when dealing with a particular transmission line) with the dimensions of voltage.

The corresponding expression for the current I can be found by substituting the result above into equation (3-12).

$$I = -\frac{1}{Z}\frac{dE}{dx} = \frac{\sqrt{ZY}}{Z}A_1 e^{-\sqrt{ZY}x} - \frac{\sqrt{ZY}}{Z}A_2 e^{\sqrt{ZY}x}$$

$$= \frac{1}{\sqrt{Z/Y}}(A_1 e^{-\sqrt{ZY}x} - A_2 e^{\sqrt{ZY}x}) \qquad (3\text{-}16)$$

The quantity $\sqrt{Z/Y}$, which has the units of ohms, is called the *characteristic impedance* (Z_0) of the line:

$$Z_0 = \sqrt{\frac{Z}{Y}} = \sqrt{\frac{R + j\omega L}{G + j\omega C}} \qquad (3\text{-}17)$$

We can note that for a lossless line (containing only L and C) the characteristic impedance is real and equal to

$$Z_0 = \sqrt{\frac{j\omega L}{j\omega C}} = \sqrt{\frac{L}{C}} \qquad \text{ohms} \qquad (3\text{-}18)$$

$$(R = G = 0)$$

It is this Z_0 which was given together with the several types of transmission lines in Section 1-2.

In general, for a lossless line it can be seen that Z_0:

1. Does not depend upon the line length.
2. Does not depend upon the line termination.
3. Depends only upon the spacing and size of the conductors and the type of dielectric used.

For a lossy line, Z_0 becomes complex and also starts to depend upon the conductor and dielectric losses (R and G).

Referring again to equations (3-15) and (3-16), the quantity $\sqrt{ZY}$ is seen to govern the manner in which the voltage and current vary with the position in the line; it governs the way in which the waves are propagated. It is given the name *propagation constant* and denoted by the symbol γ (the lowercase gamma in the Greek alphabet):

$$\gamma = \sqrt{ZY} = \sqrt{(R + j\omega L)(G + j\omega C)} \qquad (3\text{-}19)$$

The propagation constant is usually a complex number. The real part is given the symbol α (lowercase alpha in the Greek alphabet) and it determines the way in which the waves attenuate as they travel. This real part α is given the name *attenuation constant*. The imaginary part is given the symbol β (lowercase beta) and is found to determine the phase variation of the waves as they travel. For this reason it is called the *phase constant*. Hence,

$$\gamma = \alpha + j\beta \qquad (3\text{-}20)$$

where α = attenuation constant (nepers/unit length)
β = phase constant (radians/unit length)

EXAMPLE 3-3

Consider a typical open-wire transmission line which has constants of:

$$R = 14 \, \Omega/\text{mi}$$
$$L = 4.6 \, \text{mH/mi}$$
$$C = 0.01 \, \mu\text{F/mi}$$
$$G = 0.3 \times 10^{-6} \, \text{S/mi}$$

and operates at a frequency of 1000 Hz. We have, from equation (3-17),

$$Z_0 = \sqrt{\frac{R + j\omega L}{G + j\omega C}}$$

$$R + j\omega L = 14 + j28.9 \, \Omega/\text{mi} = 32.1\underline{/64.2^\circ} \, \Omega/\text{mi}$$

$$G + j\omega C = (0.3 + j62.8) \times 10^{-6} = 62.8 \times 10^{-6}\underline{/89.7^\circ} \, \text{S/mi}$$

$$Z_0 = \sqrt{\frac{32.1\underline{/64.2^\circ}}{62.8 \times 10^{-6}\underline{/89.7^\circ}}} = 715\underline{/-12.8^\circ} \, \Omega$$

$$\gamma = \sqrt{ZY} = \sqrt{32.1\underline{/64.2^\circ} \; 62.8 \times 10^{-6}\underline{/89.7^\circ}}$$

$$= 0.0449\underline{/77^\circ} = 0.01 + j.0438 \text{ per mile}$$

Therefore, the attenuation constant is

$$\alpha = 0.01 \, \text{Np/mi}$$

and the phase constant is

$$\beta = 0.0438 \, \text{rad/mi}$$

In Section 3-4 we will relate nepers to decibels.

3-3 MATCHED TRANSMISSION LINE

The general solution for the voltage along a uniform transmission line has been derived and is restated here for convenience.

$$E(x) = A_1 e^{-(\alpha + j\beta)x} + A_2 e^{+(\alpha + j\beta)x} \tag{3-15}$$

Let us now consider an infinitely long line. The second term in the general voltage equation [also for the corresponding current equation (3-16)] would tend to infinity as x is increased because of the term $e^{\alpha x}$. Since this means that the voltage and current goes to infinity for an extremely long line, a physical impossibility from an energy standpoint, the second term must be dropped or A_2 must be zero. Thus,

$$E(x) = A_1 e^{-\gamma x}$$

To obtain a value for A_1, we can use the boundary condition at the source terminals where we have labeled the voltage as E_s. From Fig. 3-5(a), $E = E_s$ at x equal to 0. Substituting this into the preceding equation, we obtain

$$E(0) = E_s = A_1 e^{-\gamma 0} = A_1$$

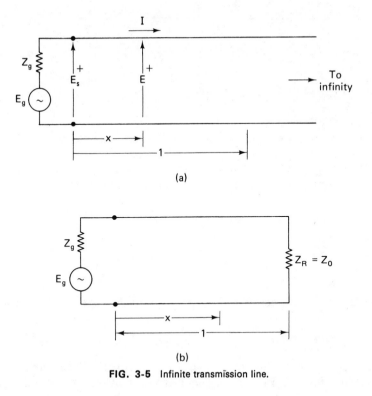

FIG. 3-5 Infinite transmission line.

Therefore,

$$A_1 = E_s \quad \text{and} \quad E(x) = E_s e^{-\gamma x} \tag{3-21}$$

The corresponding equation for the current can be found from equation (3-16) by setting $A_1 = E_s$.

$$I(x) = \frac{E_s}{Z_0} e^{-\gamma x} = \frac{E(x)}{Z_0} \tag{3-22}$$

From the last two equations the input impedance anywhere along an infinite line is seen to be

$$\frac{E(x)}{I(x)} = Z(x) = Z_0 \tag{3-23}$$

Suppose now that we cut the line at a point $x = 1$, as shown in Fig. 3-5(b). Since the line before being cut saw an impedance of Z_0, it would not electrically notice any difference if we terminated the cut line by Z_0. The finite section of line will behave as though it were an infinite line and the current and voltages would appear as given in solutions (3-21) and (3-22).

A transmission line terminated in its characteristic impedance (Z_0) is called a *matched line*. Such a line is also called a *flat line*, a *nonresonant line*, or a *properly terminated line*. The input voltage to the line (E_s) must still be solved. This is done

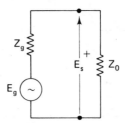

FIG. 3-6 Equivalent sending-end circuit for a matched line.

by noting that the input impedance to a matched line is equal to Z_0. Therefore, the equivalent sending-end circuit is that shown in Fig. 3-6, which allows us to solve for E_s.

$$E_s = E_g \times \frac{Z_0}{Z_g + Z_0} \qquad (3\text{-}24)$$

Incident Traveling Wave

Let us now briefly consider the expression for the voltage on a matched line:

$$E(x) = E_s e^{-(\alpha + j\beta)x}$$
$$= E_s e^{-\alpha x} e^{-j\beta x} \qquad (3\text{-}25)$$

The factor $e^{-\alpha x}$ indicates an exponential decrease in the voltage amplitude as one moves toward the load. Figure 3-7 gives a typical plot for $e^{-\alpha x}$, where the rate of

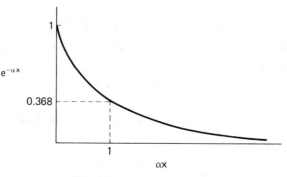

FIG. 3-7 Exponential function.

decay depends upon the attenuation constant α. The factor $e^{-j\beta x}$ indicates a progressively increasing phase lag as x increases (Fig. 3-8). This factor has a magnitude of unity and a phase angle of $-\beta x$ radians. From this we can conclude that the voltage along a lossy matched transmission line decays in amplitude as one moves toward the load. It also has a uniform progressive phase lag associated with it.

To see more clearly that this expression represents a travelling wave, we must convert it to its instantaneous form. This can be accomplished by employing equa-

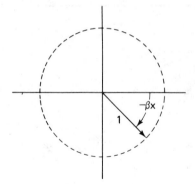

FIG. 3-8 Rotating phasor $e^{-j\beta x}$.

tion (3-4). If we take E_s to be real and an rms value, the instantaneous voltage along the line is given by

$$e = \text{Re} \left(\sqrt{2} E_s e^{-\alpha x} e^{-j\beta x} e^{j\omega t} \right)$$

$$= \sqrt{2} E_s e^{-\alpha x} \text{ Re} \left[e^{j(\omega t - \beta x)} \right]$$

$$= \sqrt{2} E_s e^{-\alpha x} \cos(\omega t - \beta x) \qquad (3\text{-}26)$$

Since the term $e^{-\alpha x}$ merely contributes to a decay in the waveform amplitude, we will temporarily assume no losses (i.e., $\alpha = 0$).

For a lossless line, equation (3-26) takes the form

$$e = \sqrt{2} E_s \cos(\omega t - \beta x) \qquad (3\text{-}27)$$

We can now plot this expression graphically for various values of time. Table 3-1 gives the values of the expression $\cos(\omega t - \beta x)$ as a function of βx for the times $\omega t = 0$, $\omega t = \pi/4$, and $\omega t = \pi/2$. These are, in turn, plotted in Fig. 3-9. From Fig. 3-9 we can note that as time increases, the wave moves toward the $+x$ direction or toward the receiving end of the transmission line. In general, we can conclude that

TABLE 3-1 Values of $\cos(\omega t - \beta x)$

	$\cos(\omega t - \beta x)$		
βx	$\omega t = 0$	$\omega t = \pi/4$	$\omega t = \pi/2$
0	1	0.707	0
$\pi/8$	0.924	0.924	0.383
$\pi/4$	0.707	1.000	0.707
$3\pi/8$	0.383	0.924	0.924
$\pi/2$	0	0.707	1.00
$5\pi/8$	-0.383	0.383	0.924
$3\pi/4$	-0.707	0.000	0.707
$7\pi/8$	-0.924	-0.383	0.383
π	-1	-0.707	0.00

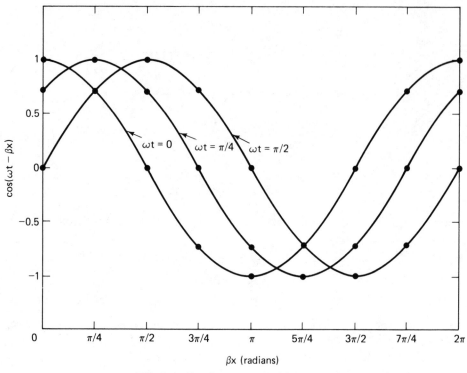

FIG. 3-9 Traveling wave on a lossless line.

an expression of the form $e^{-\gamma Z}$ represents a wave traveling in the plus Z direction. If the exponential is positive (i.e., $e^{+\gamma Z}$), the expression represents a wave traveling in the $-Z$ direction.

For a lossy matched line the general waveform observed as one moves along the line for three different instances of time is sketched in Fig. 3-10. In transmission-line theory, the general expressions for the voltage and current are [see equations

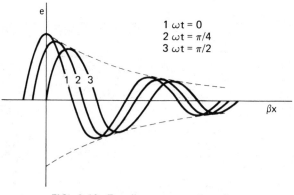

FIG. 3-10 Traveling wave on a lossy line.

(3-15) and (3-16)]

$$E(x) = A_1 e^{-\gamma x} + A_2 e^{\gamma x} \tag{3-28}$$

$$I(x) = \frac{A_1 e^{-\gamma x}}{Z_0} - \frac{A_2 e^{\gamma x}}{Z_0} \tag{3-29}$$

where the factors containing $e^{-\gamma x}$ are the incident waves (E^+ or I^+) and the factors containing the $e^{\gamma x}$ are the reflected waves (E^- and I^-). The matched line has only an incident wave on it.

To determine the phase velocity of the wave (how fast the *wave travels*), let us imagine an observer following a constant phase point on the wave; that is, ($\beta x - \omega t$) = a constant in equation (3-27). The phase velocity can be obtained by taking the derivative of this expression with respect to time:

$$\frac{d}{dt}(\beta x) - \frac{d}{dt}\omega t = \frac{d}{dt} \text{ constant} \qquad \text{where } \beta \text{ and } \omega \text{ are constants}$$

$$\beta \frac{dx}{dt} - \omega = 0$$

Since dx/dt is the velocity we desire, the expression for the phase velocity v_p becomes

$$v_p = \frac{dx}{dt} = \frac{\omega}{\beta} \qquad \text{length/second} \tag{3-30}$$

As an example, let us obtain the phase velocity on a lossless line. In this case $\alpha = 0$ and $\gamma = \alpha + j\beta = \sqrt{j\omega L \cdot j\omega C} = j\omega\sqrt{LC}$, or

$$\beta = \omega\sqrt{LC} \tag{3-31}$$

The phase velocity on such a line is

$$v_p = \frac{\omega}{\beta} = \frac{\omega}{\omega\sqrt{LC}} = \frac{1}{\sqrt{LC}} \tag{3-32}$$

For a lossless line, the velocity of energy propagation is given by the same expression.

Before considering a concrete example, let us also establish the relationship between the wavelength and the phase constant on a transmission line. Since the wavelength (λ) is the distance a wave travels in one cycle (an equivalent angle of 2π radians) and the phase constant (β) indicates the phase shift per unit length on the line, then

$$\beta\lambda = 2\pi \text{ rad}$$

or

$$\lambda = \frac{2\pi}{\beta} \tag{3-33}$$

This can also be expressed in terms of the phase velocity v_p, where substituting equation (3-30) for the wavelength becomes

$$\lambda = \frac{2\pi}{\omega}v_p = \frac{v_p}{f} \tag{3-34}$$

EXAMPLE 3-4

Consider a 100-mi open-wire flat telephone line which has the same characteristics as those given in Example 3-3. The frequency of the generator is 1 kHz and it has an internal impedance of 600 Ω. Find the:

(a) Sending-end current.
(b) Sending-end voltage.
(c) Sending-end power.
(d) Receiving-end current.
(e) Receiving-end voltage.
(f) Receiving-end power.
(g) Power loss (in dB).

The open-circuited generator voltage is $10\underline{/0°}$ volts rms.

Solution: From the solution of Example 3-3,

$$Z_0 = 715\underline{/-12.8°}\ \Omega = 697 - j158\ \Omega$$

$$\gamma = \alpha + j\beta = 0.01 + j.0438 \text{ per mile}$$

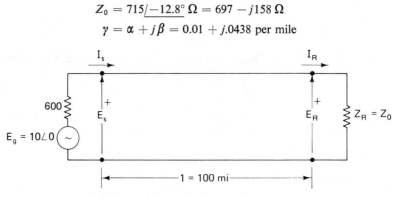

FIG. 3-11 Transmission line used in example.

For a matched line, $Z_s = Z_0$. Therefore,

$$I_s = \frac{E_g}{Z_g + Z_0} = \frac{10\underline{/0°}}{600 + 697 - j158} = \frac{10\underline{/0°}}{1307\underline{/-6.96°}}$$

$$= 7.65\underline{/6.96°}\ \text{mA}$$

$$E_s = I_s Z_s = 7.65\underline{/6.96°} \times 10^{-3} \times 715\underline{/-12.8°}$$

$$= 5.47\underline{/-5.8°}\ \text{V}$$

$$P_s = |E_s||I_s| \times \text{power factor}$$

$$= 5.47 \times 7.65 \times 10^{-3} \cos (6.96° + 5.8°)$$

$$= 41.8 \cos 12.8°\ \text{mW}$$

$$= 40.8\ \text{mW}$$

Another method of computing P_s is by using the expression

$$P_s = |I_s|^2 R_s$$

$$= (7.65 \times 10^{-3})^2 \times 697$$

$$= 40.8\ \text{mW}$$

For the receiving-end voltage, we use equation (3-21), where x is made equal to the length of line (1 = 100 mi).

$$E_R = E_s e^{-\alpha l} e^{-j\beta l}$$
$$= 5.47\underline{/-5.8°}\ e^{-0.01 \times 100} e^{-j.0438 \times 100}$$
$$= 5.47 e^{-1.0} \underline{/-5.8°} - 4.38 \text{ rad}$$
$$= 5.47 \times 0.368 \underline{/-5.8° - 251°}$$
$$= 2.02 \underline{/-257°}$$
$$I_R = \frac{E_R}{Z_R} = \frac{2.02\underline{/-257°}}{715\underline{/-12.8°}} = 2.83\underline{/-244°} \text{ mA}$$
$$P_R = |I_R|^2 R_R = (2.83 \times 10^{-3})^2 \times 697$$
$$= 5.58 \text{ mW}$$

The transmission-line loss is $10 \log (40.8/5.58) = 8.7$ dB.

EXAMPLE 3-5

Find the wavelength of the signal on the line in Example 3-4 and the length of line in terms of wavelengths.

Solution: From equation (3-33),

$$\lambda = \frac{2\pi}{\beta} = \frac{2\pi}{0.0438} = 143.5 \text{ mi}$$

$$l \text{ (in wavelengths)} = \frac{100}{143.5} = 0.697\lambda$$

3-4   THE NEPER AND THE DECIBEL

In this section the relation of the neper to the decibel is discussed. If we consider two points (x_1 and x_2) on a matched transmission line (Fig. 3-12), the magnitudes of the voltages in terms of the sending-end voltage are, respectively [from equation (3-25)],

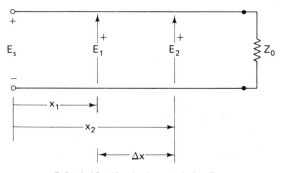

FIG. 3-12 Matched transmission line.

$$|E_1| = |E_s| e^{-\alpha x_1} \qquad (3\text{-}35)$$

$$|E_2| = |E_s| e^{-\alpha x_2}$$

The ratio of the voltage magnitudes at the two locations are

$$\frac{|E_2|}{|E_1|} = \frac{|E_s| e^{-\alpha x_2}}{|E_s| e^{-\alpha x_1}} = e^{-\alpha(x_2 - x_1)} = e^{-\alpha \Delta x} \qquad (3\text{-}36)$$

where $\alpha \Delta x$ is the total attenuation in nepers between the two points x_1 and x_2.

The expression for the total number of nepers of gain (it will be a negative gain or a loss) can be obtained by taking the natural logarithm of the last equation.

$$\text{number of nepers gain} = \alpha \Delta x = -\ln \frac{|E_2|}{|E_1|} \qquad (3\text{-}37)$$

where $|E_2| < |E_1|$.

To obtain the decibel in terms of nepers, we must go to the basic definition of the decibel:

$$\text{no. dB} = 10 \log_{10} \frac{P_2}{P_1} \qquad (3\text{-}38)$$

where P_2 = power at point 2
P_1 = power at point 1

In terms of the voltages at locations x_1 and x_1, $P_2 = |E_2|^2/Z_0$ and $P_1 = |E_1|^2/Z_0$, where the Z_0 is assumed to be real. Thus,

$$\text{no. dB gain} = 10 \log \left(\frac{|E_2|}{|E_1|} \right)^2 = 20 \log \frac{|E_2|}{|E_1|}$$

Substituting equation (3-36) for the ratio $|E_2/E_1|$, we obtain

$$\text{no. dB gain} = 20 \log e^{-\alpha \Delta x}$$

$$= -\alpha \Delta x \, 20 \log e$$

$$= -\alpha \Delta x \, 8.686$$

Since $-\alpha \Delta x$ represents the total loss in nepers, a loss of 1 Np will represent 8.686 dB. Hence,

$$1 \text{ Np} = 8.686 \text{ dB} \qquad (3\text{-}39)$$

EXAMPLE 3-6

Find the transmission-line loss in nepers and decibels of the transmission line given in Example 3-4.

Solution:

$$\text{no. nepers loss} = \alpha l = 0.01 \times 100$$

$$= 1.0 \text{ Np}$$

$$\text{no. decibels loss} = 8.686 \times 1.0 \text{ Np}$$

$$= 8.7 \text{ dB}$$

In order to pass a complex signal through a transmission line without causing any distortion, the attenuation of the signal containing many sinusoidal components should be constant with frequency, and the phase velocity for the various frequency components should also be held constant. In this way, the form of the received signal will be the same as the transmitted wave, only somewhat attenuated and delayed in time. If the phase velocity must be a constant with frequency, the phase constant β must be proportional to the frequency, as can be observed from equation (3-32). Therefore, for distortionless transmission

$$\alpha = \text{constant}$$

$$\beta = \omega \times \text{constant}$$

Now, in general, the propagation constant for a transmission line is given by

$$\gamma = \sqrt{(R + j\omega L)(G + j\omega C)}$$

If we have a lossless line where $R = G = 0$, we have distortionless transmission as $\alpha = 0$ and $\beta = \omega\sqrt{LC}$. Also at very high frequencies, the line tends to be distortionless, where it can be shown that α and β can be approximated by

$$\alpha \approx \frac{R}{2}\sqrt{\frac{C}{L}} + \frac{G}{2}\sqrt{\frac{L}{C}} \qquad (3\text{-}40a)$$

$$\beta \approx \omega\sqrt{LC} \qquad (3\text{-}40b)$$

In the lower voice-frequency range, however, these approximations do not hold and the line is generally not distortionless.

A distortionless line can still be obtained, however, if the constants of the line are forced to bear the following relationship:

$$\frac{R}{L} = \frac{G}{C} \qquad (3\text{-}41)$$

Under this condition, by substituting in the equation for the propagation constant,

$$\alpha = \sqrt{RG}$$

$$\beta = \omega\sqrt{LC}$$

and

$$v_p = \frac{1}{\sqrt{LC}}$$

For a representative telephone cable, the ratio R/L is usually much greater, by a magnitude of a thousand or so, than G/C. To obtain the distortionless condition, then, G could be increased, but this would greatly increase the attenuation; R could be reduced, resulting in large-diameter conductors. C could also be reduced by a larger spacing between conductors, but this has practical limits. The common method generally employed is to increase L by inductive loading. Typically, loading coils are spaced about 1 mi apart in a voice frequency cable circuit.

Ideally, the inductance should be distributed along the line, but this scheme, although used in transatlantic cables, is expensive. Employing lumped inductances at fixed points along the line has the disadvantage of causing the circuit to become a low-pass filter. For this reason, a compromise is made in the magnitude of the inductances inserted and the spacing along the line. A certain amount of distortion is usually tolerated and the inductance does not arrive at the values required by the ratio given in (3-41).

The advantages of loading are best illustrated by the following example.

EXAMPLE OF LOADING

A 19-gauge pair in a toll cable used for VF (voice frequency) has the following characteristics:

$$R = 86 \ \Omega/\text{loop mile}$$

$$L = 1 \ \text{mH/loop mile}$$

$$G = 1.4 \ \mu\text{S/loop mile}$$

$$C = 0.062 \ \mu\text{F/loop mile}$$

(The loop mile takes into consideration that there is a wire pair for one complete circuit.) For ideal loading,

$$\frac{R}{L} = \frac{G}{C}$$

In this case,

$$\frac{R}{L} = 86,000$$

$$\frac{G}{C} = 22.6$$

The amount of inductance (L_L) that must be added for distortionless transmissions for every mile should be

$$\frac{R}{L_L + L} = 22.6$$

$$L_L = \frac{86}{22.6} - L = 3.8 \ \text{H/mi} - 1 \ \text{mH/mi}$$

$$\approx 3.8 \ \text{H/mi}$$

If this amount of loading is added, a cutoff frequency of around 650 Hz is obtained, which is much too low for VF. We shall show that by adding only 150 mH/mi, the results are quite acceptable.

By using these values for a frequency of 300 Hz (the low end of the VF band) and for 3300 Hz (the high end of the VF band), one obtains the following results:

Non loaded

300 Hz $\alpha = 0.0704 \ \text{Np/mi}$

$$v_p = \frac{2\pi f}{\beta} = 2.65 \times 10^4 \ \text{mi/s}$$

$$3300 \text{ Hz} \qquad \alpha = 0.211 \text{ Np/mi}$$

$$v_p = 7.89 \times 10^4 \text{ mi/s}$$

Loaded (150 mH added)

$$300 \text{ Hz} \qquad \alpha = 0.0284 \text{ Np/mi}$$

$$v_p = 1.02 \times 10^4 \text{ mi/s}$$

$$3300 \text{ Hz} \qquad \alpha = 0.0286 \text{ Np/mi}$$

$$v_p = 1.03 \times 10^4 \text{ mi/s}$$

By inspecting the results, the following conclusions can be drawn.

1.   The large increase in loss (about 200% for frequencies near 3200 Hz as compared to 300 Hz) does not occur when the cable is loaded. α is almost constant and there is no attenuation distortion.

2.   The considerable higher velocity of the high-frequency components is practically eliminated when the pair is loaded. The high- and low-frequency components travel at about the same speed and there is no delay distortion.

3.   The reduction in loss, especially at the high end of the VF range after loading is quite pronounced.

4.   The velocity of propagation is considerably reduced with loading. This is a disadvantage when echos are received from mismatched loads, as these longer time delays make the echos more annoying.

5.   Although not shown in these calculations, loading does tend to cut off the high end of the VF band.

6.   The characteristic impedance under ideal loading becomes real and equal to $\sqrt{R/G}$. This results in a resistive load when matching is generally advantageous.

7.   The loading coils increase the magnetic field strength, which results in more crosstalk between adjacent lines.

3-6 MISMATCHED TRANSMISSION LINE

We must now consider a mismatched line which has present, in addition to an incident wave, a reflected wave. It may be recalled that a reflected wave takes the form of $e^{+\gamma x}$ (a wave traveling in the negative x direction) or equivalently $e^{-\gamma d}$ (a wave traveling in the plus d direction or from the load to the source). The nomenclature that will be used is shown in Fig. 3-13.

Our aim is to obtain the total voltage (and current) at some point x on the line in terms of E_s (the total sending-end voltage). The procedure is to sum the incident wave present at point x and the reflected wave present at point x. This is done initially in terms of the incident sending-end voltage $[E^+(0)]$ which is later converted to terms of E_s. *Note:* $E_s = E^+(0) + E^-(0)$.

The total voltage at x is given by

$$E(x) = E^+(x) + E^-(x) \qquad (3-42)$$

where $E^+(x)$ = incident voltage at point x

$E^-(x)$ = reflected voltage at point x

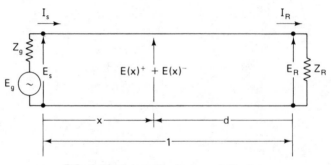

FIG. 3-13 Schematic of a transmission line.

$E^+(x)$ can be written in the form

$$E^+(x) = E^+(0)e^{-\gamma x} \tag{3-43}$$

where $E^+(0)$ is the incident sending-end voltage. Expression (3-43) is similar to equation (3-21), but now the incident sending-end voltage is no longer identical with E_s, since a reflected voltage is also present.

The *reflected voltage* can be obtained by following the incident wave to the load, undergoing a reflection, and returning to point x (distance d from the load). The incident voltage at the load is

$$E^+(l) = E^+(0)e^{-\gamma l} \tag{3-44}$$

The reflected voltage at the load is equal to

$$E^-(l) = \Gamma_R E^+(l)$$
$$= \Gamma_R E^+(0)e^{-\gamma l} \tag{3-45}$$

by substitution of equation (3-44). Γ_R is the *load reflection coefficient*, which is defined as the ratio of the reflected to incident voltage. It is related to the load impedance by equation (2-7). Looking back from the load, this reflected voltage when arriving at point d (or x) is equal to

$$E^-(d) = E^-(l)e^{-\gamma d}$$

since

$$E^-(d) = E^-(x) \qquad \text{at the same point}$$
$$= E^-(x) = E^-(l)e^{-\gamma(l-x)}$$

Substituting equation (3-45) for $E^-(l)$, we obtain

$$E^-(x) = \Gamma_R E^+(0)e^{-\gamma l}e^{-\gamma(l-x)}$$
$$= E^+(0)\Gamma_R e^{-\gamma(2l-x)} \tag{3-46}$$

The total voltage anywhere along the line is obtained by substituting equations (3-43) and (3-46) into equation (3-42):

$$E(x) = E^+(0)[e^{-\gamma x} + \Gamma_R e^{-\gamma(2l-x)}] \tag{3-47}$$

Since $E^+(0)$ in this expression cannot be directly obtained or easily measured, it will be expressed in terms of the sending-end voltage E_s. At $x = 0$,

$$E(0) = E_s = E^+(0)(1 + \Gamma_R e^{-2\gamma l})$$

Therefore,

$$E^+(0) = \frac{E_S}{1 + \Gamma_R e^{-2\gamma l}} \tag{3-48}$$

Equation (3-47) then becomes

$$E(x) = E_S \frac{e^{-\gamma x} + \Gamma_R e^{-\gamma(2l-x)}}{1 + \Gamma_R e^{-2\gamma l}}$$

Multiplying both numerator and denominator by $e^{\gamma l}$, we obtain

$$E(x) = E_S \frac{e^{\gamma(l-x)} + \Gamma_R e^{-\gamma(l-x)}}{e^{\gamma l} + \Gamma_R e^{-\gamma l}} \tag{3-49}$$

The corresponding expression for the current can also be found by noting that the incident and reflected currents see the characteristic impedance of the line; that is,

$$I^+(x) = \frac{E^+(x)}{Z_0}$$

and

$$I^-(x) = -\frac{E^-(x)}{Z_0}$$

The negative sign appearing because of the convention that positive currents flow toward the load.

The total current can then be found from equations (3-43) and (3-46) or (3-47) and is equal to

$$I(x) = I^+(x) + I^-(x) = \frac{E^+(0)}{Z_0} e^{-\gamma x} - \frac{E^+(0)}{Z_0} \Gamma_R e^{-\gamma(2l-x)}$$

$$I(x) = \frac{E^+(0)}{Z_0} [e^{-\gamma x} - \Gamma_R e^{-\gamma(2l-x)}]$$

Again substituting equation (3-48) for $E^+(0)$:

$$I(x) = \frac{E_s}{Z_0} \frac{e^{-\gamma x} - \Gamma_R e^{-\gamma(2l-x)}}{1 + \Gamma_R e^{-2\gamma l}}$$

$$= \frac{E_s}{Z_0} \frac{e^{\gamma(l-x)} - \Gamma_R e^{-\gamma(l-x)}}{e^{\gamma l} + \Gamma_R e^{-\gamma l}} \tag{3-50}$$

To obtain E_s theoretically, the input impedance to the line must be known. The input impedance is simply obtained by dividing the voltage by the current at $x = 0$:

$$Z_s = \frac{E(0)}{I(0)} = Z_0 \frac{e^{\gamma l} + \Gamma_R e^{-\gamma l}}{e^{\gamma l} - \Gamma_R e^{-\gamma l}} \tag{3-51}$$

In order to simplify the mathematics, yet still be able to obtain a good picture of what happens on a resonant or mismatched line, the lossless line will be considered.

($R = G = 0$ and hence $\alpha = 0$.) Later, a few additional comments will be made regarding the lossy line.

To reduce equations (3-49), (3-50), and (3-51) to the lossless case, α should be set to zero. All exponential terms in these equations can now be converted to sine & cosine functions. Thus, when $\alpha = 0$,

$$e^{\gamma(l-x)} = e^{j\beta(l-x)} = \cos\beta(l-x) + j\sin\beta(l-x)$$
$$e^{-\gamma(l-x)} = e^{-j\beta(l-x)} = \cos\beta(l-x) - j\sin\beta(l-x)$$

and similarly for $e^{\gamma l}$ and $e^{-\gamma l}$.

If these substitutions are made in the equations previously mentioned and if the further substitution for $\Gamma_R = (Z_R - Z_0)/(Z_R + Z_0)$ is made, the voltage, current, and impedance equations result in

$$E(x) = E_S \frac{Z_R \cos\beta(l-x) + jZ_0 \sin\beta(l-x)}{Z_R \cos\beta l + jZ_0 \sin\beta l} \tag{3-52}$$

$$I(x) = \frac{E_S}{Z_0} \frac{Z_0 \cos\beta(l-x) + jZ_R \sin\beta(l-x)}{Z_R \cos\beta l + jZ_0 \sin\beta l} \tag{3-53}$$

$$Z_s = Z_0 \frac{Z_R \cos\beta l + jZ_0 \sin\beta l}{Z_0 \cos\beta l + jZ_R \sin\beta l} \tag{3-54}$$

As the matched line has already been dealt with, we shall consider two extreme conditions which surprisingly enough are frequently employed. These are the (a) short-circuited line, and the (b) open-circuited line.

3-7 SHORT-CIRCUITED LOSSLESS TRANSMISSION LINE

From equation (3-52) it is found that the voltage at the load end of the line E_R is

$$E_R = E(l) = \frac{E_S Z_R}{Z_R \cos\beta l + jZ_0 \sin\beta l} \tag{3-55}$$

thus

$$E_S = \frac{E_R}{Z_R}(Z_R \cos\beta l + jZ_0 \sin\beta l) \tag{3-56}$$

If we consider a length of line $d = l - x$ (the distance referred to from the load end), equation (3-52) can be rewritten as

$$E(d) = E_S \left[\frac{Z_R \cos\beta d + jZ_0 \sin\beta d}{Z_R \cos\beta l + jZ_0 \sin\beta l} \right]$$

substituting equation (3-56) for E_S we obtain

$$E(d) = E_R \left(\cos\beta d + j\frac{Z_0}{Z_R} \sin\beta d \right)$$
$$= E_R \cos\beta d + jI_R Z_0 \sin\beta d \tag{3-57}$$

In analogous fashion the corresponding equation for the current is

$$I(d) = I_R \cos\beta d + j\frac{E_R}{Z_0} \sin\beta d \tag{3-58}$$

For the short-circuited line, $Z_R = 0$ and $E_R = 0$; therefore, equations (3-57) and (3-58) become for the shorted case:

$$E_{sc}(d) = iI_R Z_0 \sin \beta d = jI_R Z_0 \sin \frac{2\pi}{\lambda} d \qquad (3\text{-}59)$$

$$I_{sc}(d) = I_R \cos \beta d = I_R \cos \frac{2\pi}{\lambda} d \qquad (3\text{-}60)$$

From the last equations we note that the voltages and current are always 90° out of phase. This should be expected, since no power is absorbed in a lossless shorted line and therefore the power factor should be zero ($\cos 90° = 0$). It can also be observed that the voltage and current vary sinusoidally in magnitude as one moves along the transmission line. The voltage is zero at the short and becomes a maximum at a distance of $\lambda/4$ from the load. The current, however, is a maximum at the short and is a minimum at a distance of $\lambda/4$ from the load. The resulting *standing-wave* pattern for the current and voltage is shown in Fig. 3-14. The phase angle of the voltage and current is also shown in Fig. 3-14. It should be recalled that the sign of the sine function changes when going from the second to the third

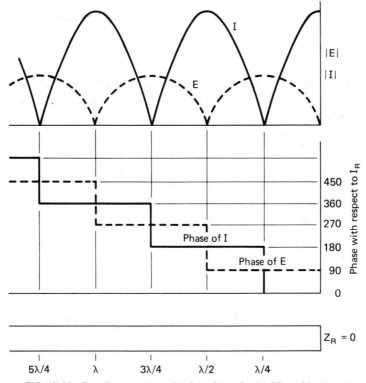

FIG. 3-14 Standing waves on a lossless short-circuited line with phase information.

quadrant and from the fourth to the first quadrant. For the cosine function the sign changes between the first and second quadrants and between the third and fourth quadrants. This is shown in Fig. 3-15. The minimum voltage points are often called the *nodes* on the line. From Fig. 3-14 we can observe that the distance between the nodal points is one-half of a wavelength. This phenomenon is frequently used to determine the frequency of a signal at high frequencies. (For an open-air coaxial line, $f = c/\lambda$.)

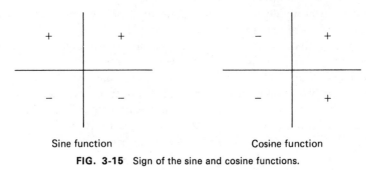

Sine function   Cosine function

FIG. 3-15   Sign of the sine and cosine functions.

The shorted-line and the matched line are frequently used in measurement techniques to obtain a varying voltage magnitude of fixed phase and a varying voltage phase of fixed magnitude. This can be briefly stated as:

1. A shorted transmission line can be used to obtain a voltage of *varying magnitude* and constant phase by moving along the line between two adjacent nodes.

2. A matched transmission line ($E = E_s e^{-j\beta x} = E_s \underline{/-\beta x}$) can be used to obtain a voltage of fixed magnitude but *varying phase*.

The impedance seen looking into a length of short-circuited transmission line can be obtained from equation (3-54) or by dividing the voltage expression (3-59) by the current expression (3-60). By performing the latter, we obtain

$$Z_{sc}(d) = \frac{E_{sc}(d)}{I_{sc}(d)}$$

$$= \frac{jI_R Z_0 \sin\left[(2\pi/\lambda)\,d\right]}{I_R \cos\left[(2\pi/\lambda)\,d\right]} = jZ_0 \tan \frac{2\pi}{\lambda}\,d \qquad (3\text{-}61)$$

From equation (3-61) we can observe that the impedance of a lossless short-circuit transmission line is purely reactive. The input impedance (or rather reactance) of the shorted line as a function of its length is shown in Fig. 3-16.

From the equation (3-61) or Fig. 3-16, it can be seen that a shorted lossless line is either capacitive or inductive, depending upon the length of the line. If $d < \lambda/4$, the line appears inductive, and the closer that d approaches $\lambda/4$, the larger

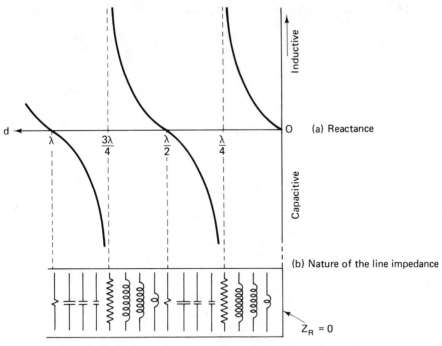

FIG. 3-16 Variation of reactance along a lossless short-circuited line.

this inductance becomes. If the line is greater than $\lambda/4$ long but less than $\lambda/2$ long, the line appears to be capacitive. It should be noted that the impedance repeats itself every half-wavelength along the line.

Applications of Shorted Transmission Lines

Some typical applications of a resonant line will now be considered.

Coaxial Insulating Support

At a quarter wavelength from the short, the stub support appears as an open circuit (Fig. 3-17). It often results in less reflection than a dielectric supporting bead. Since the physical length of the shorted stub is fixed, it is inherently a narrow-band device.

Choke Joint

A useful form of coupling waveguides is shown in Fig. 3-18. The L-shaped channel may be regarded as a half-wave transmission line. At point C, which is a quarter-wavelength from the closed end B, the impedance is high and a poor contact at C can be accepted. In certain circumstances a gap of several millimeters may be tolerated. The impedance as seen at point A is very low.

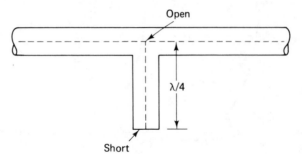

FIG. 3-17 Coaxial insulating support.

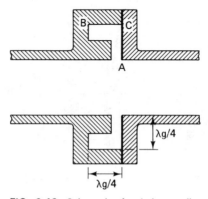

FIG. 3-18 Schematic of a choke coupling.

Baluns

The *balun* is a device used to interconnect an unbalanced line to a balanced line without disturbing the equilibrium condition on either. At low frequencies, a common type of transformer is often employed (see Fig. 3-19). Here the primary side is unbalanced since one lead has a low impedance with respect to ground, whereas on the secondary side, both leads have identical impedances with respect to ground (balanced). One such transformer made by Channel Master (CM Model 7280) uses a ferrite voltage step-up transformer (2:1) which can be used in the frequency range 50–890 MHz. It also has the feature of impedance matching a 75-Ω unbalanced line to a 300-Ω balanced circuit.

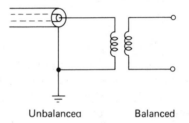

Unbalanced   Balanced

FIG. 3-19 Low-frequency transformer used as a balun.

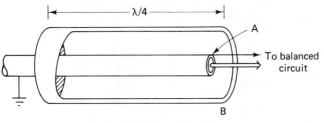

FIG. 3-20   "Bazooka" balun.

At high frequencies, where a coaxial cable having the outer sheath at ground potential is frequently employed, one often must convert to a balanced two-wire line or a dipole antenna. One such device is shown in Fig. 3-20. It consists of a sleeve concentric with the outer sheath of a coaxial line, with the one end shorted to the sheath. The sleeve and the cable sheath form a new quarter-wavelength coaxial line, which is short-circuited at one end. This produces a high impedance between points A and B, where B is at ground (at dc both points A and B are at ground). Point A is now isolated from ground and can form one side of the balanced circuit. The inner conductor was always isolated from ground. The advantage of this balun over the one discussed next is that the standing wave that results is in the sleeve and therefore does not radiate.

Another balun, similar in theory to the one discussed previously, is shown in Fig. 3-21 feeding a dipole antenna. Again, point B has a high impedance with

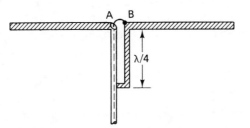

FIG. 3-21   Quarter-wavelength balun.

respect to ground, owing to the short to open-circuit conversion of a $\lambda/4$ shorted transmission line. A matched transmission line can also be used to form a balun. Such a device, called a *half-wave-line balun*, is shown in Fig. 3-22. If one moves along a matched transmission line a half-wavelength, a phase shift of 180° occurs (polarity reversal). This can be shown by looking at equation (3-25) for a lossless line.

$$E(x) = E_s e^{-j\beta x} = E_s e^{-j(2\pi x/\lambda)}$$

since $\beta = 2\pi/\lambda$. If $x = \lambda/2$,

$$\beta x = \frac{2\pi}{\lambda} \frac{\lambda}{2} = \pi \text{ rad or } 180°$$

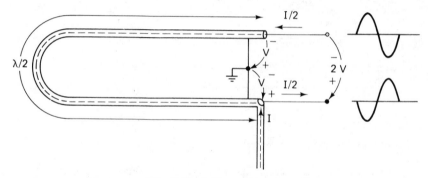

FIG. 3-22 Half-wave-line balun.

The half-wave-line balun has associated with it a $4:1$ impedance transformation. This can be shown by noting that the output voltage is $2V$, where V is the voltage at the input terminals. The output current is $I/2$. Thus,

$$Z_{out} = \frac{2V}{I/2} = 4\frac{V}{I} = 4Z_{in}$$

If a 300 balanced and matched load is connected to such a balun, the impedance seen at the input is $300/4 = 75 \ \Omega$.

Coaxial Resonator

Another application of shorted resonant lines shown in Fig. 3-23. In this example concentric coaxial resonators are employed as tuned circuits which are adjusted, depending on the desired frequency of operation. The lighthouse tube, as such, has

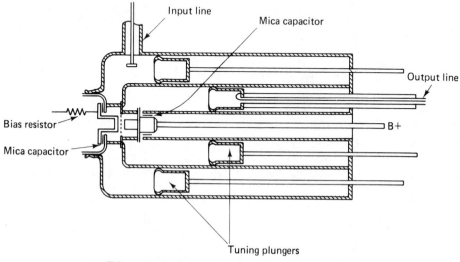

FIG. 3-23 Typical amplifier circuit for a lighthouse tube.

small interelectrode spacing so that the transit line is minimized. These tubes have been made to oscillate at frequencies exceeding 6 GHz, but generally they are reliable only up to 3 GHz. More will be said about resonant circuits in Section 3-9.

3-8 OPEN-CIRCUITED LOSSLESS TRANSMISSION LINE

If a similar development is used for the open-circuited line as for the short-circuited line, one obtains for the input impedance the expression

$$Z_{oc}(d) = -jZ_0 \text{ cotangent } \frac{2\pi}{\lambda} d \qquad (3\text{-}62)$$

This is plotted in Fig. 3-24. For the open-circuited line, if the line length is less than

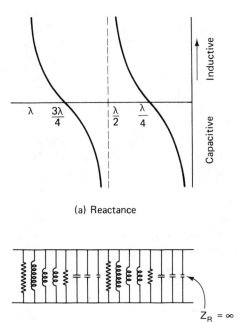

(a) Reactance

(b) Nature of the line impedance

FIG. 3-24 Variation of reactance along a lossless open-circuited line.

$\lambda/4$, the input impedance appears capacitive. The line can be conceived of as two parallel plates for $d < \lambda/4$. If the line is $\lambda/4$ long, the input appears as a short. From a length of $\lambda/4$ to $\lambda/2$ it appears inductive.

The open-circuited line is not all that frequently used, since fringing of the fields occur at the open end and the location of the open cannot be accurately determined. Also, at sufficiently high frequencies, the lines start radiating off the open end and the load begins to appear resistive. The voltage and current along an open-circuited line appears as shown in Fig. 3-25.

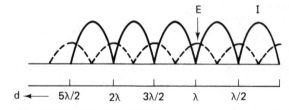

FIG. 3-25 Voltage and current standing waves on an open-circuited line.

3-9 RESONANT CIRCUITS

At sufficiently high frequencies, short sections of transmission line with completely reflecting terminations frequently replace the conventional inductor coil. The distributed capacitance between the turns of a coil greatly reduces the impedance as the frequency is increased, causing the coil to behave in a very complex manner. Also, because of radiation, losses become needlessly high. Capacitors are similarly replaced, owing to the distributed inductance of the leads, but their range of usefulness can be considerably extended by employing small plates of simple geometry.

Since resonant circuits are made up of parallel and series capacitors and inductors, at high frequencies, shorted and open-circuited transmission lines frequently make up the resonant circuit. A quarter-wavelength-long short-circuited transmission line, for instance, appears as an open circuit and can be considered to be a parallel resonant (antiresonant) circuit. Similarly, a short-circuited line one half-wavelength long appears as a short circuit and behaves electrically as a series resonant circuit. To obtain a better understanding of the electrical behavior of such a circuit, let us consider a short-circuited line which has a fixed length of $l = \lambda_0/4$ at a frequency f_0 (Fig. 3-26). The input impedance of this line will then be observed as the frequency is varied.

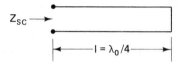

FIG. 3-26 Line of fixed length $\lambda_0/4$.

The input impedance looking into a shorted line is equal to [from equation (3-61)]

$$Z_{sc} = jZ_0 \tan \beta l \qquad (3-63)$$

since

$$\beta = \frac{\omega}{v_p} = \frac{2\pi f}{v_p} = \frac{2\pi f}{f_0 \lambda_0}$$

Equation (3-63) can be rewritten as

$$Z_{sc} = jZ_0 \tan \frac{2\pi f}{f_0 \lambda_0} \frac{\lambda_0}{4}$$

$$= jZ_0 \tan \frac{\pi f}{2f_0} \qquad (3\text{-}64)$$

A plot of this expression is shown in Fig. 3-27. Since at f_0 the impedance goes toward infinity, this reminds one very much of the antiresonance condition of a lumped LC parallel circuit. On the other hand, at $2f_0$, the impedance of the line goes to zero, which closely resembles the behavior of a lumped LC series circuit around a resonant frequency of $2f_0$. Superimposed on Fig. 3-27 is the behavior of

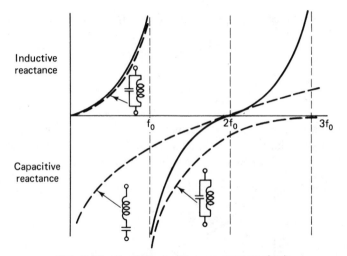

FIG. 3-27   Short-circuited line as a resonant circuit.

a typical lumped parallel circuit antiresonant at f_0 and of a lumped series circuit resonant at $2f_0$. This plot indicates that at f_0, the $\lambda_0/4$ shorted transmission line behaves very similarly to the parallel resonant LC circuit. This is repetitive at $3f_0$, $5f_0$, and so on. At $2f_0$, $4f_0$, . . . , the line behaves similarly to the series resonant LC case. The higher-order modes are used, as in the case of the resonant structure shown in Fig. 3-23, where the plunger may strike the tube in the $\lambda/4$ mode, but these have the disadvantage of raising the circuit Q and thereby decreasing the modulation bandwidth.

The actual electrical equivalent to one of the resonating structures shown in Fig. 3-23 is given in Fig. 3-28. The capacitor C represents the respective interelectrode capacitances of the high-frequency tube. The system is antiresonant when the impedance looking into the length of transmission line is a reactance equal and

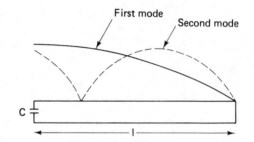

FIG. 3-28 Equivalent circuit of a tube-coaxial line resonator.

opposite to that of the capacitors reactance. The input impedance looking into the shorted line of length l can be expressed as

$$Z_{sc} = jZ_0 \tan \frac{\omega}{v_p} l$$

Equating this reactance with the negative of the capacitive reactance, we obtain

$$jZ_0 \tan \frac{\omega}{v_p} l = -\frac{-j}{\omega c}$$

or

$$Z_0 \tan \frac{\omega l}{v_p} = \frac{1}{\omega c} \qquad (3\text{-}65)$$

The antiresonant frequencies are obtained from the intersections of the function $Z_0 \tan (\omega l / v_p)$ with the hyperbola $1/\omega c$ as shown in Fig. 3-29. It should be noted that the antiresonant frequencies are not harmonically related.

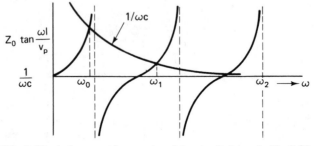

FIG. 3-29 Antiresonant frequencies of the circuit shown in Fig. 3-28.

The Q and Bandwidth of Resonant Lines

The resonant circuits discussed thus far are idealized in the sense that losses in the lines have not been taken into account. In the practical case there are conductor and dielectric losses which cause the input impedance at antiresonance to be finite rather than infinite, and the input impedance at series resonance to take on a nonzero resistive value.

At antiresonance, it can be shown that the impedance is resistive and given by

$$Z_{max} = \frac{Z_0}{\alpha l} \qquad \alpha l \ll 1 \qquad (3\text{-}66)$$

Likewise, the impedance at series resonance has a small resistive value of

$$Z_{min} = Z_0 \alpha l \qquad \alpha l \ll 1 \qquad (3\text{-}67)$$

The expression (3-40a) can be used for α when evaluating these maximum and minimum impedance values.

If the line is highly lossy, these expressions do not hold and both Z_{max} and Z_{min} tend toward Z_0 for $\alpha l \gg 1$. Such a lossy structure should not be considered to be a resonant circuit. The two parameters often encountered when dealing with resonant circuits are the quality factor (Q) and the bandwidth (BW) near the resonant or antiresonant points. The *bandwidth*, or the degree of sharpness of the resonant impedance curve, is taken to be the number of hertz between the half-power frequencies or the $1/\sqrt{2}$ voltage frequencies. The *quality factor* is a measure of the lossyness of the line as related to the energy that can be stored in the line. The general definition of Q is given by the expression

$$Q = \omega \times \frac{\text{stored energy}}{\text{power dissipated}} \qquad (3\text{-}68)$$

This definition can be applied to any resonant structure, such as lumped circuits, resonant cavities, and mechanical resonant systems. If, for instance, the stored energy in a system is held constant and a lossy component is added, such as a large resistance inserted in parallel with a tank circuit, the Q decreases. The bandwidth of a system is related to Q by the expression

$$BW = \frac{f_0}{Q} \qquad (3\text{-}69)$$

As can be observed by the last expression, large Q's or low-loss systems result in small bandwidths.

The Q for a low-loss short-circuited transmission line ($\alpha l \ll 1$) is independent of the number of quarter-wavelengths of the resonant circuit and given by

$$Q = \frac{\beta}{2\alpha} \qquad \alpha l \ll 1 \qquad (3\text{-}70)$$

When substituting for α in expression (3-40a), we obtain

$$Q = \frac{\omega_0 LC}{RC + LG} \qquad (3\text{-}71)$$

If an air-dielectric line having a $G = 0$ is used, as is often the real case, Q is simplified to

$$Q = \frac{\omega_0 L}{R} \qquad G = 0 \qquad (3\text{-}72)$$

Although for lumped resonant circuits a Q of 100 may be considered high, Q's of several thousand can readily be obtained with transmission lines. One disadvantage of a high Q is that the buildup and decay times under transient conditions are slow. It takes a number of oscillations before a waveform reaches a new steady-state condition.

3-10 VOLTAGE STANDING-WAVE RATIO

A new term can now be defined, the *voltage standing-wave ratio*, or VSWR. The relative magnitude of the reflected wave on a low-loss line is generally expressed by means of the voltage standing-wave ratio, which is defined as

$$\text{VSWR} = \frac{|E_{\text{max}}|}{|E_{\text{min}}|} \tag{3-73}$$

where E_{max} is the rms (or peak) voltage at the highest point in the standing-wave pattern and E_{min} is the rms (or peak) value shown in Fig. 3-30. The VSWR is often

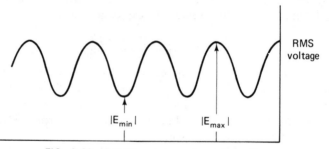

FIG. 3-30 Standing wave on a transmission line.

expressed in terms of decibels, which is related to the VSWR by the relation

$$\text{VSWR (dB)} = 20 \log_{10} \text{VSWR} \tag{3-74}$$

The VSWR is widely used because it is one of the more easily measured quantities on a line. A flat line, which has no reflected wave, has $|E_{\text{max}}| = |E_{\text{min}}|$ and a VSWR of unity. The ratio becomes larger without limit as complete reflection is approached. The VSWR cannot be directly measured when the standing-wave pattern changes its form markedly from one loop to another, as in the case of high losses.

The VSWR is related to the magnitude of the reflection coefficient by noting that maximum voltage occurs on the line when the incident and reflected waves add in phase or when

$$|E_{\text{max}}| = |E^+| + |E^-|$$

where $E^+ + E^-$ are defined in Section 3-6. The minimum voltage occurs at the point where the incident and reflected waves are collinear but 180° out of phase or when

$$|E_{\text{min}}| = |E^+| - |E^-|$$

Thus,

$$\text{VSWR} = \frac{|E_{max}|}{|E_{min}|} = \frac{1 + |E^-/E^+|}{1 - |E^-/E^+|} = \frac{1 + |\Gamma|}{1 - |\Gamma|} \tag{3-75}$$

In equation (3-75) the voltage reflection coefficient is defined as the ratio of the reflected voltage to the incident voltage at any point on the line or

$$\Gamma = \frac{E^-}{E^+} \tag{3-76}$$

EXAMPLES

The VSWR will now be found for some typical loads.

(a) *Short-circuited line* ($Z_R = 0$)

$$\Gamma = \frac{Z_R - Z_0}{Z_R + Z_0} = -1$$

$$|\Gamma| = 1$$

Therefore,

$$\text{VSWR} = \frac{1 + 1}{1 - 1} = \infty$$

(b) *Open-circuited line* ($Z_R \longrightarrow \infty$)

$$\Gamma = 1$$

$$\text{VSWR} = \infty$$

(c) *Matched line* ($Z_R = Z_0$)

$$\Gamma = 0$$

$$\text{VSWR} = \frac{1 + 0}{1 - 0} = 1$$

(d) $Z_R = 2Z_0$

$$\Gamma = \frac{2Z_0 - Z_0}{2Z_0 + Z_0} = \frac{Z_0}{3Z_0} = 1/3$$

$$\text{VSWR} = \frac{1 + 1/3}{1 - 1/3} = \frac{4/3}{2/3} = 2$$

From this, one can gather that the minimum VSWR is 1. This must be true, of course, since E_{max} must be less then E_{min} if VSWR is to be less than 1, which is a contradiction.

3-11 NOTE ON LOSSY LINES

The discussion so far has been concerned with the ideal case of a lossless line. Although this can never be achieved in practise, the losses on lines at sufficiently high frequencies are so small that the predictions based on a lossless line are reasonably close to the actual results. We will now modify the graphical sketches of voltage and current on a transmission line to correspond with a lossy line. The corresponding equations will not be derived, not because it is so difficult but because it will be easier to do this by the graphical approach given later. Figure 3-31 shows the voltage and current along a lossy line which is shorted. Note that the VSWR

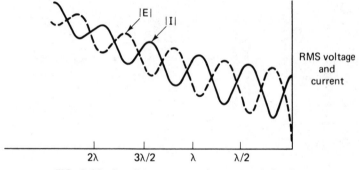

FIG. 3-31 Standing waves on a lossy transmission line.

becomes less as the sending end is approached. The same result would occur if a pad or attenuator is inserted in a mismatched line.

PROBLEMS

3-1. The characteristic impedance (Z_0) of a lossy transmission line depends upon these (*check all correct answers*):
(a) Length of the line.
(b) Frequency of the applied signal.
(c) Dielectric constant of the insulation.
(d) Load at the end of the line.
(e) Separation and size of the wires.

3-2. Obtain the rms phasors of the following:
(a) $i = 100 \cos(\omega t - \pi/8)$
(b) $v = 70 \cos(\omega t + \pi)$

3-3. Obtain the expression for the instantaneous values of the following:
(a) $I = 10 \,\underline{/\pi/4}$ A rms.
(b) $V = 100 \,\underline{/-\pi/2}$ V peak.

3-4. Obtain the rms phasor of the waveform shown on the oscilloscope face.
$\qquad$ Vertical amplifier: 10 V/cm
$\qquad$ Horizontal deflection: 1 m s/cm

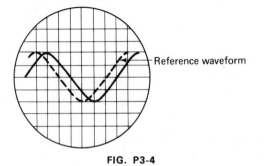

Reference waveform

FIG. P3-4

3-5. Convert the following to rectangular and polar form.

(a) $e^{-0.01} =$

    $e^{-2}\quad =$

    $e^{2}\quad =$

(b) $e^{-j1} =$

    $e^{-j.01} =$

    $e^{-j\pi/4} =$

    $e^{j2}\quad =$

(c) $e^{-1-j^1}\quad =$

    $e^{-2-j2}\quad =$

    $e^{1+j^1}\quad =$

    $e^{-1-j\pi/4} =$

3-6. An open-wire line has the following constants:

$$R = 14\ \Omega/\text{mi} \qquad G = 0$$
$$L = 5\ \text{mH/mi} \qquad f = 1\ \text{kHz}$$
$$C = 0.02\ \mu\text{F/mi}$$

Obtain:

(a) Z_0

(b) γ

(c) α

(d) B

3-7. An open-wire telephone line has the following line constants:

$$R = 4.04\ \Omega/\text{mi}$$
$$L = 4\ \text{mH/mi}$$
$$G = 0$$
$$C = 0.01\ \mu\text{F/mi}$$
$$f = 1\ \text{kHz}$$

Obtain the following for the line:

(a) Characteristic impedance.

(b) Propagation constant.

(c) Attenuation constant.

(d) Phase constant.

(e) Phase velocity.

3-8. (a) What is the relationship between the phase constant and wavelength?

(b) If a 100-mi matched transmission line has an α of 0.007 Np/mi, what will the total loss of the line be (in dB)?

3-9. A certain telephone line has the following electrical characteristics:

$$\gamma = 0.007 + j.04/\text{mi}$$
$$Z_0 = 600 - j100\ \Omega$$

If $10\underline{/0°}$ V (rms) is applied to the sending end (generator impedance $= 500\ \Omega$), determine the following for a 200-mi-long matched line:

(a) Z_S

(b) I_S

(c) E_S

(d) P_S

(e) E_R

(f) P_R

(g) Loss (in dB)

3-10. Describe what the following equations represent:

(a) $E = K_1 e^{-\gamma Z} \qquad (\gamma = \alpha + j\beta)$

(b) $E = K_2 e^{\gamma Z}$

3-11. A certain telephone cable has the following electrical characteristics:

$$R = 40 \ \Omega/\text{mi}$$
$$L = 1.1 \ \text{mH/mi}$$
$$G = \text{negligible}$$
$$C = 0.062 \ \mu\text{F/mi}$$

Loading coils are added which provide an additional inductance of 30 mH/mi, as well as an additional resistance of 8 Ω/mi. Obtain the attenuation constant and phase velocities at frequencies of 300 Hz and 3300 Hz. If the coil spacing is to be $\lambda/6$ at a frequency of 3300 Hz, find the physical distance between coils.

3-12. If $i = \text{Re} \ (I_m \ e^{j\omega t})$ and $v = \text{Re} \ (V_m \ e^{j\omega t})$:
(a) Find the relationship between the phasors I_m and V_m across an inductor where

$$v = L \frac{di}{dt}$$

(b) Find the relationship between the phasors I_m and V_m across a capacitor where

$$i = C \frac{dv}{dt}$$

3-13. Derive equation (3-58) from equation (3-50) assuming a lossless transmission line.

3-14. A short-circuited coaxial transmission line has a $Z_0 = 52 + j0$ ohms and a propagation constant of $0.0 + j9.4$/m. What is the input impedance if the line length is 6 cm?

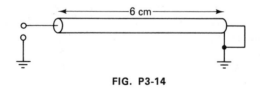

FIG. P3-14

3-15. If the line of Problem 3-14 is $3\lambda/8$ long, what would be its input impedance?

3-16. (a) What is the input impedance of an 11-cm length of shorted lossless open-wire line? ($f = 3$ GHz.)
(b) What is the input impedance of the same shorted line as in part (a) if the length is 10 cm? 2.5 cm? ($f = 3$ GHz.)

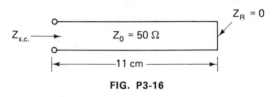

FIG. P3-16

3-17. (a) What is a balun?
(b) Define VSWR.

3-18. A lossless open-circuited transmission line has an output voltage E_R of 10 V. If $\beta = 1.256$ rad/m, and the line length is 50 m, what is the voltage amplitude at a distance of 1 m from the load end? At 1.25 m from the load? At 2.5 m?

3-19. The input and output voltages on a 50-mi-long matched transmission line are measured to be $10\,\underline{|0°}\ V_{rms}$ and $5\,\underline{|57.30°}\ V_{rms}$, respectively. Obtain the following for the line shown.

(a) Z_S (d) α

(b) I_S (e) β

(c) E_S

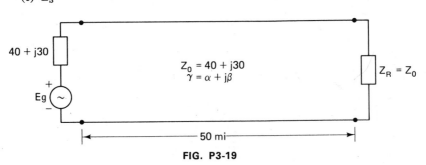

FIG. P3-19

3-20.

For the circuits shown, determine the following:

(a) Z_a (c) Z_c

(b) Z_b (d) Z_d

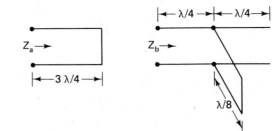

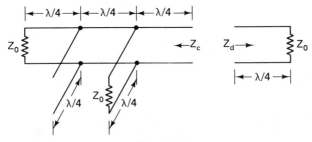

FIG. P3-20

3-21. An open-circuited coaxial line is to be used as an antiresonant circuit as shown in Fig. 3-28. If $C = 2$ pF, determine the required length of 50-Ω line required for antiresonance at 250 MHz. $\epsilon_r = 1$

3-22. A short-circuited quarter-wave section of 50-Ω Andrew HELIAX Type HJ10 (see Fig. 1-6) is to be used as an antiresonant circuit at 500 MHz. Compute the Q, the BW, and the sending-end impedance at antiresonance.

four

THE SMITH CHART

4-1 INTRODUCTION

In order to reduce the amount of labor when calculating characteristics of transmission lines, various charts have been developed. If, for instance, the input impedance of a line is desired, equation (3-51) could be worked out with a slide rule or calculator, or one of the charts could be employed. Naturally, the accuracy obtained by the slide rule or calculator routine is much higher than that obtained with the graphical technique; nevertheless, the latter method is sufficiently close for most cases. The chart can be used for both lossy and lossless lines, and they facilitate many calculations which can otherwise become quite tedious. Since the Smith chart is the one most commonly used, we will employ it. It basically is an impedance chart, giving a graphical indication of the impedance of a transmission line as one moves along it.

Since the derivation of the Smith chart is given in Appendix B, we will begin to become familiar with its use by doing several problems. Before doing so, the following comments should be made:

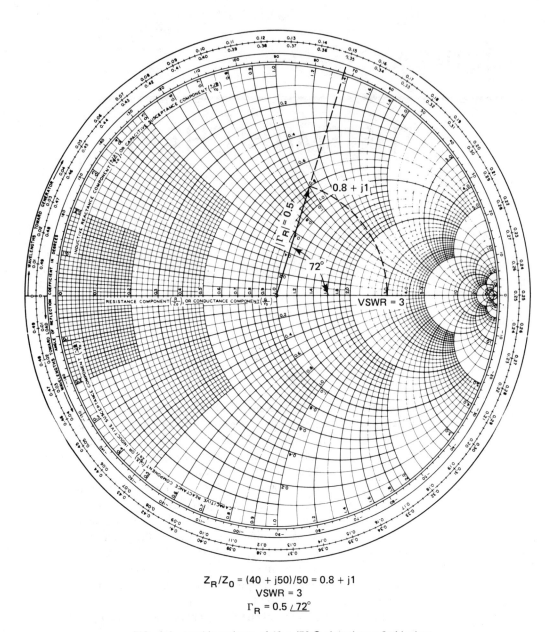

$$Z_R/Z_0 = (40 + j50)/50 = 0.8 + j1$$
$$\text{VSWR} = 3$$
$$\Gamma_R = 0.5 \,\underline{/72°}$$

FIG. 4-1 Load impedance of 40 + j50 Ω plotted on a Smith chart.

1. The Smith chart we are using is a normalized chart where all impedances are divided by Z_0. If, for instance, a 50-Ω transmission line is terminated in a load impedance of $100 + j50\ \Omega$, this load impedance is plotted as $(100 + j50)/50 = 2 + j1$ on the Smith chart. There are Smith charts of 50 Ω, 75 Ω, and so on, but these have the disadvantage of being employable only on lines of corresponding characteristic impedance.

2. The Smith charts that are used have included a protractor (measured in degrees) and a distance scale (in wavelengths) around the circumference of the chart. The wavelength scale has no absolute value; it only aids in determining how far to rotate around the chart as one advances along the transmission line.

3. The reflection coefficient can be plotted on the Smith chart by radially going from the center of the chart toward the outside a distance equal to its magnitude (the scale is linear, having a maximum value of 1 when at the chart circumference) at an angle equal to the phase of the reflection coefficient. This is shown in Fig. B-3 of Appendix B. Along a lossless transmission line, the phase angle of the reflection coefficient varies, while its magnitude remains fixed. Therefore, as one moves down a lossless line, the reflection coefficient rotates around the Smith chart at a fixed radial distance from the chart center.

4. Since the VSWR is equivalent to the real quantity $(Z/Z_0)_{max}$ (see Appendix B), it can readily be determined on the Smith chart by noting the intercept of the reflection coefficient locus on the zero reactance line for $R/Z_0 > 1$. Figure 4-1 shows the plot of the load impedance $40 + j50$ on the Smith chart when connected to a 50-Ω transmission line. The corresponding voltage reflection coefficient and VSWR are also shown.

4-2 PROBLEM-SOLVING PROCEDURES

Let us now proceed to do some typical problems on the Smith chart.

EXAMPLE 4-1

Find the input impedance of a lossless line a third of a wavelength long, which is terminated in an impedance of $150 + j60\ \Omega$ as shown in Fig. 4-2.

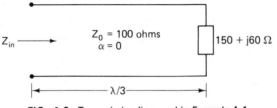

FIG. 4-2 Transmission line used in Example 4-1.

Solution: The solution may be presented in three simple steps:

1. Plot the normalized load impedance

$$\frac{Z_R}{Z_0} = \frac{150 + j60}{100} = 1.5 + j.6$$

on the Smith chart, shown as point A on Fig. 4-3.

2. Construct a radial line from the center of the Smith chart to point A and note on the "Wavelengths Toward Generator" scale the value 0.198. Since we must find the impedance at the point $\lambda/3$ toward the generator from the load locations, 0.333λ must be added to the 0.198λ value, as previously noted. This results in a value of $(0.500 + 0.031)\lambda = 0.531\lambda$. Since the wavelength scale around the Smith chart stops at 0.500, one must continue to travel 0.031λ beyond this point or to the 0.031 point on the wavelengths toward generator scale. Construct a radial line from the center through this 0.031 reading.

3. Swing an arc about the center from point A on the first radial line until it intersects at point B on the second radial line. (Since α is zero, only the phase angle of the reflection coefficient changes.) Read the normalized input impedance at point B as $(0.55 + j0.15)$. Multiplying by Z_0, the input impedance is found to be $55 + j15\ \Omega$.

We can find the load reflection coefficient either analytically [equation (2-7)] or graphically (Fig. 4-3). At the receiving end,

$$\Gamma_R = \frac{Z_R - Z_0}{Z_R + Z_0} = \frac{150 + j60 - 100}{150 + j60 + 100} = \frac{5 + j6}{25 + j6}$$

$$= \frac{7.81\underline{/50.2°}}{25.7\underline{/13.5°}} = 0.304\underline{/36.7°}$$

This is seen to be the same value as that obtained from the Smith chart.

At the input end, the reflection coefficient can be obtained from equation (4-3), which is derived later.

$$\Gamma = \Gamma_R e^{-2\gamma d} = \Gamma_R e^{-j2\beta d} \qquad \text{if } \alpha = 0$$

$$= 0.304\underline{/36.7°}\left|-2 \times \frac{2\pi}{\lambda} \times \lambda/3 \text{ rad}\right.$$

$$= 0.304\left/36.7° - \frac{4\pi}{3} \times \frac{180°}{\pi}\right.$$

$$= 0.304\underline{/-203.3°} = 0.304\underline{/156.7°}$$

This agrees with the value obtained directly from the Smith chart. On this particular Smith chart the magnitude of the reflection coefficient can be read from the linear reflection coefficient scale on the bottom of the sheet. The simplest method of doing this is to use a pair of dividers. Spread it the distance from the center of the chart to point B, and then bring it down to the reflection coefficient scale and note its magnitude.

As we have noted earlier, $(Z/Z_0)_{\max} = $ VSWR, and so the VSWR can be

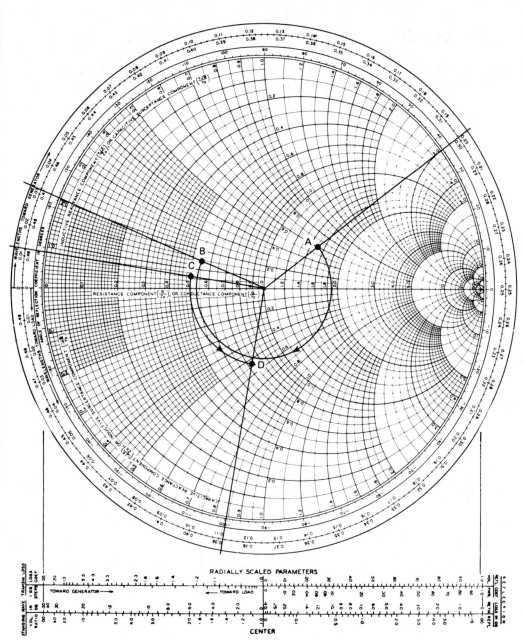

FIG. 4-3 Smith chart for Examples 4-1 and 4-2.

77

found by locating the maximum normalized impedance point. This occurs to the right of the chart (reactance of zero) and reads 1.88.

EXAMPLE 4-2

Using the same Smith chart (Fig. 4-3), find the load impedance at the end of a $\lambda/8$ line if the sending-end impedance is $50 + j7\,\Omega$. $Z_0 = 100\,\Omega$. This type of problem is frequently encountered in practice, since generally the load is electrically somewhat remote from the measuring point at high frequencies.

Solution: Plot the normalized impedance $(50 + j7)/100 = 0.5 + j.07$ on the Smith chart. Label this point C. Then rotate on a constant $|\Gamma|$ circle or constant VSWR circle

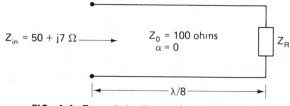

FIG. 4-4 Transmission line used in Example 4-2.

$\lambda/8 = 0.125\lambda$ toward the load. This point, labeled D, will be the normalized load impedance.

$$\frac{Z_R}{Z_0} = 0.73 - j.53$$

$$Z_R = 73 - j53\,\Omega$$

EXAMPLE 4-3

The VSWR on a lossless line is measured to be 5, with a voltage minimum occurring $\lambda/3$ from the load. Determine the load impedance if $Z_0 = 50\,\Omega$.

Solution: The VSWR of the 5 circle is first drawn on the Smith chart of Fig. 4-6. Since the voltage minimum in the line has a corresponding current maximum, this particular

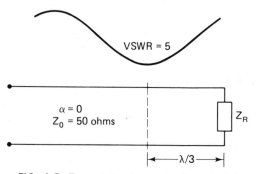

FIG. 4-5 Transmission line used in Example 4-3.

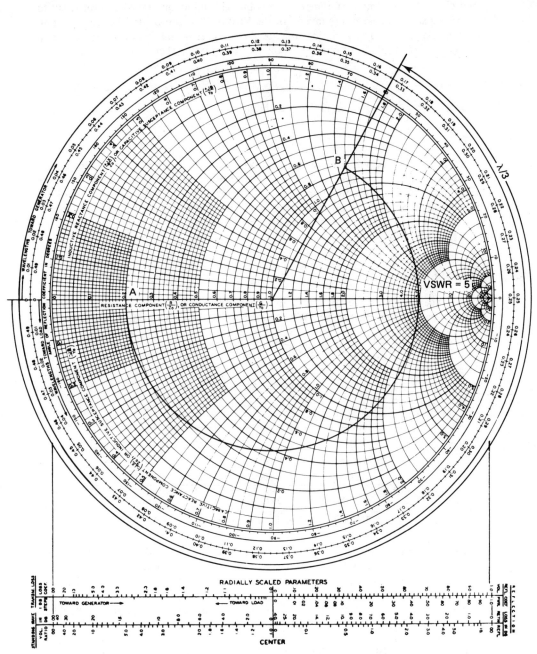

FIG. 4-6 Smith chart for Example 4-3.

point on the line is a minimum impedance point. This minimum impedance turns out to be a real quantity equal to the inverse of the VSWR. This can be derived from relation (B-3) in Appendix B, by noting that Z/Z_0 becomes a minimum when the term $|\Gamma_R|\,\underline{/\phi - 2\beta l}$ is 180° out of phase with the 1 or unity term. Then

$$\frac{Z}{Z_0} = \left(\frac{Z}{Z_0}\right)_{min} = \frac{1 - |\Gamma_R|}{1 + |\Gamma_R|}$$

Comparing this with equation (3-75), we obtain

$$\left(\frac{Z}{Z_0}\right)_{min} = \frac{1}{VSWR} \tag{4-1}$$

The minimum voltage point on the line therefore corresponds to the minimum impedance point on the Smith chart, which occurs where the VSWR circle intersects the zero reactance locus (point A). From this minimum impedance point we must rotate $\lambda/3$ toward the load to obtain the normalized load impedance (point B). The normalized load impedance is seen to be $0.77 + j1.48$. Therefore, the load impedance is

$$Z_R = 50(0.77 + j1.48) = 38.5 + j74 \ \Omega$$

EXAMPLE 4-4

Two voltage minimums are noted on a mismatched line at distances of 10 cm and 50 cm from the load. With a measured VSWR of 3, determine the load impedance (see Fig. 4-7).

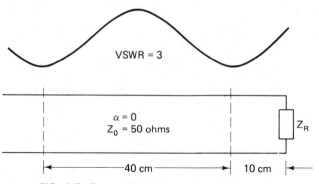

FIG. 4-7 Transmission line used in Example 4-4.

Solution: Since the impedance at a voltage minimum is also a minimum, the locations of the voltage minima can be located on the Smith chart. This occurs where the VSWR = 3 locus intercepts the zero reactance line for minimum resistance (see Fig. 4-8). In the theory section (refer, for instance, to Fig. 3-14) we have noted that the distance between voltage nodes or voltage minimum points is one-half wavelength. Thus, for this problem

$$\lambda/2 = 40 \text{ cm} \quad \text{or} \quad \lambda = 80 \text{ cm}$$

To obtain the normalized load impedance, therefore, we must rotate 10/80 wavelength, or $\lambda/8$, from the impedance minimum point toward the load. This results in a normalized load impedance of $0.6 - j.8$, or a load impedance of $50(0.6 - j.8) = 30 - j40 \ \Omega$.

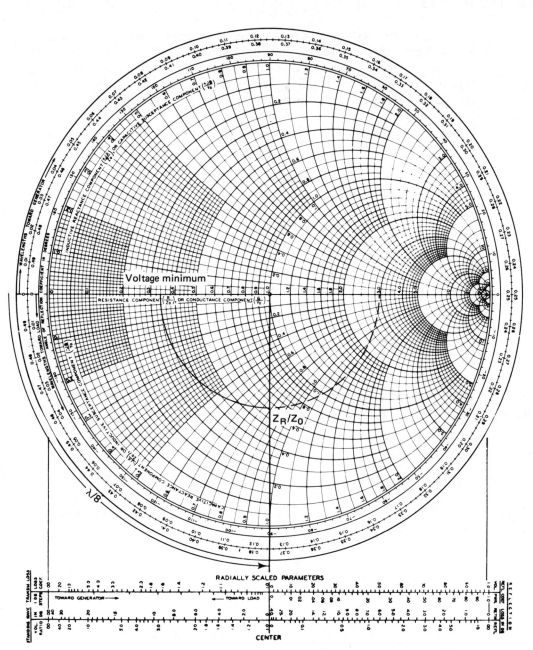

FIG. 4-8 Smith chart for Example 4-4.

4-3 LOSSY LINES

Up to now when performing any calculations we have pretty well concentrated on the lossless line, since such a simplified treatment normally gives a sufficiently clear picture of the electrical behavior of a line. The general expressions (3-49) to (3-51) can be used to determine the impedance, voltage, and current anywhere along a line, which often involves a lot of labor, particularly in the case of the lossy line. The Smith chart can also be employed to obtain the impedance along a lossy line. The less-often-required current and voltage must still be obtained by using the appropriate equations with the suitable values substituted in for the generator voltage and internal impedance.

When dealing with lossy lines on the Smith chart, one must also consider the attenuation or change in magnitude of the reflection coefficient as one moves along the line. To obtain the general expression for the reflection coefficient, which is defined as the ratio of the reflected voltage (E^-) to the incident voltage (E^+), we will make use of equations (3-43) and (3-46). Substituting these relationships in for E^- and E^+, we obtain

$$\Gamma = \frac{E^-}{E^+} = \frac{E^+(0)\Gamma_R e^{-\gamma(2l-x)}}{E^+(0)e^{-\gamma x}}$$

$$= \Gamma_R e^{-2\gamma(l-x)} = \Gamma_R e^{-2\gamma d} \tag{4-2}$$

Letting $\Gamma_R = |\Gamma_R|\underline{/\phi}$,

$$\Gamma = |\Gamma_R|\, e^{-2\alpha d}\, \underline{/\phi - 2\beta d} \tag{4-3}$$

Thus, as one moves along a lossy line the phase of the reflection coefficient remains identical to that of the lossless line (i.e., $\phi - 2\beta d$), but the magnitude of the reflection coefficient varies exponentially as

$$|\Gamma| = |\Gamma_R|e^{-2\alpha d} \tag{4-4}$$

As one moves toward the generator end with increasing d, the reflection coefficient decreases in magnitude. The associated VSWR also decreases [equation (3-75)].

EXAMPLE 4-5

Find the sending-end impedance of a 100-mi line having the following constants at 1000 Hz.

$$Z_0 = 685 - j92\ \Omega$$

$$\gamma = 0.00497 + j.0352$$

$$Z_R = 2000\ \Omega$$

Solution: Plot the normalized impedance:

$$\frac{Z_R}{Z_0} = \frac{2000}{685 - j92} = 2.87 + j.385 \text{ on the Smith chart}$$

The magnitude of the reflection coefficient at the load as determined from the Smith chart (Fig. 4-9) is

$$|\Gamma_R| = 0.491$$

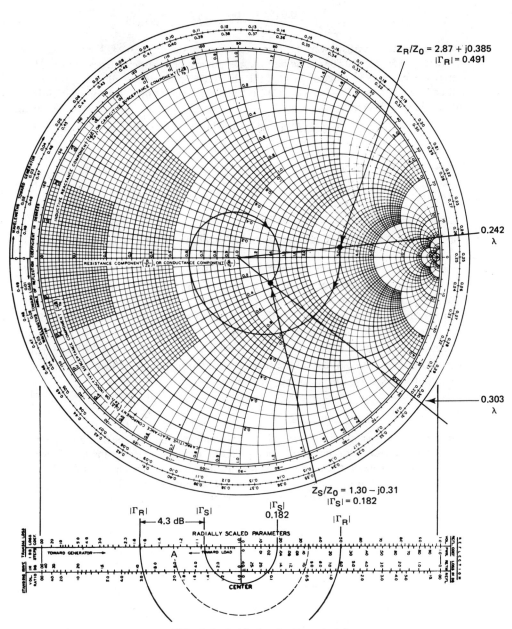

$Z_R/Z_0 = 2.87 + j0.385$
$|\Gamma_R| = 0.491$

0.242
λ

0.303
λ

$Z_S/Z_0 = 1.30 - j0.31$
$|\Gamma_S| = 0.182$

FIG. 4-9 Smith chart for Example 4-5.

To obtain the impedance at the sending end (Z_S) we must rotate 100 mi toward the generator from the load. This is equivalent to

$$\frac{l}{\lambda} = \frac{\beta}{2\pi}l = \frac{0.0352 \times 100}{2\pi} = 0.561 \text{ wavelength}$$

Since on the wavelengths toward generator scale we are starting at 0.242λ, we should stop at $(0.242 + 0.561)\lambda = 0.803\lambda$. One-half wavelength can be subtracted from this since this merely means an extra rotation around the chart. Therefore, the sending-end radial line will be located at $(0.803 - 0.5)\lambda = 0.303\lambda$. The magnitude of the reflection coefficient at the sending end becomes

$$|\Gamma_S| = |\Gamma_R|e^{-2\alpha d} = 0.491e^{-2 \times 0.00497 \times 100}$$

$$= 0.491 \times 0.37 = 0.182$$

Hence, the normalized sending-end impedance read from the Smith chart is

$$\frac{Z_S}{Z_0} = 1.30 - j.31 \quad \text{or} \quad Z_S = Z_0(1.30 - j.31)$$

$$= 861 - j332 \ \Omega$$

The VSWR at the receiving end is seen to be 2.93 while at the sending end it drops to 1.44.

On this particular Smith chart there is also a transmission loss scale which can be employed to determine the magnitude of the reflection coefficient. On this scale, the distance between each mark represents 1 dB. In the problem at hand, the magnitude of the reflection coefficient at the load is 0.491. This is marked off on the transmission loss scale (from the origin, called the center). Moving toward the generator a loss of 100 mi $\times$ 0.00497 Np/mi $\times$ 8.68 dB/Np = 4.3 dB is experienced and therefore, as noted on the scale, the magnitude of the reflection coefficient must be decreased by this amount.

EXAMPLE 4-6

The attenuation constant of a transmission line of known length can be found by measuring the input VSWR with the output shorted. As an example, consider the line shown in Fig. 4-10. The reflection coefficient at the load is unity ($|\Gamma_R| = 1$), whereas the reflection coefficient as determined from a Smith chart at the sending end is 0.333. Substituting into equation (4-4), we can obtain the attenuation constant (in Np/ft):

$$|\Gamma_S| = 0.33 = 1e^{-2\alpha 500}$$

or

$$e^{1000\alpha} = \frac{1}{0.33}$$

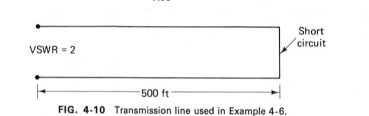

FIG. 4-10   Transmission line used in Example 4-6.

Taking the natural logarithm of both sides, we get

$$1000\alpha = \ln \frac{1}{0.33} = \ln 3 = 1.1$$

or

$$\alpha = 1.1 \times 10^{-3} \text{ Np/ft}$$
$$= 9.55 \times 10^{-3} \text{ dB/ft}$$

Alternatively, using the TRANS. LOSS scale on the Smith chart, the distance between $|\Gamma| = 1$ and $|\Gamma| = 0.33$ (Point A on Fig. 4-9) represents 4.8 steps or 4.8 dB. Thus the loss per foot is given by

$$\frac{4.8}{500} = 9.6 \times 10^{-3} \text{ dB/ft}$$

PROBLEMS

4-1. (a) If you plot a normalized load impedance on a Smith chart, how would you find:
   (i) Its reflection coefficient on the chart?
   (ii) Its VSWR on the chart?
   (b) Why do you move along a constant VSWR circle on the Smith chart when moving along a lossless transmission line?
   (c)

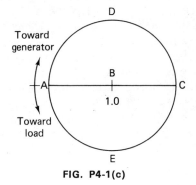

FIG. P4-1(c)

Indicate the following on the outline of the Smith chart above:
   (i) $(Z/Z_0)_{\min}$
   (ii) $(Z/Z_0)_{\max}$
   (iii) $0 + j0$

4-2. For the transmission line shown, find the:
   (a) Reflection coefficient (Γ_R) at the receiving end.
   (b) VSWR on the line.
   (c) Input impedance (Z_{in}).

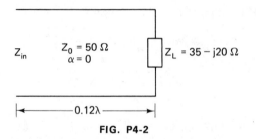

FIG. P4-2

4-3. A transmission line with a $Z_0 = 50\,\Omega$ and $\alpha = 0$ has a voltage minimum 0.42λ from the load end. The VSWR $= 2.9$. Find the:
(a) Load impedance Z_L.
(b) Reflection coefficient at the load end.

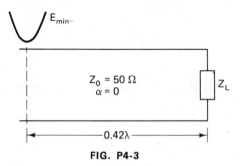

FIG. P4-3

4-4. An antenna, as a load on a transmission line, produces a VSWR of 2.8, with a voltage minimum at 0.12λ from the antenna terminals. Find the antenna impedance and the reflection coefficient at the antenna, if $Z_0 = 300\,\Omega$ for the line.

4-5. A transmission line is terminated in Z_L. Measurements indicate that the standing-wave minima are 102 cm apart, and that the last minimum is 35 cm from the load end of the line. The value of VSWR is 2.4 and $Z_0 = 250\,\Omega$.
(a) Find Z_L in terms of real and reactive components.
(b) What frequency is being transmitted ($v_p = c$)?

4-6. 75 miles of open-wire transmission line is terminated with a load impedance of $1200 + j0\,\Omega$. The line has a $Z_0 = 600 - j100\,\Omega$, and a propagation constant of $\gamma = 0.005\,\text{Np/mi} + j.05\,\text{rad/mi}$ at 1000 Hz. Find:
(a) Γ_R
(b) Γ at sending end
(c) Z_{in}

4-7.

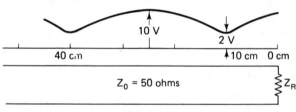

FIG. P4-7

The voltage maxima and minima are measured as indicated above on a transmission line ($Z_0 = 50\ \Omega$) which is terminated by Z_R. Calculate Z_R and Γ_R.

4-8. The loss in a transmission line can be determined by shorting the receiving end and noting the VSWR at the sending end. On a particular transmission line of 10 mi in length, it is found that the VSWR at the sending end is 2.0 when the output terminals are shorted. Find the:

(a) Magnitude of the reflection coefficient at the receiving and sending end.

(b) Loss of the line (in Np). ($N.B.: |\Gamma| = |\Gamma_R| e^{-2\alpha d}$.)

(c) Loss of the line (in dB).

4-9. Determine the length of a 50-Ω short-circuited line required to produce a reactance of $+j100\ \Omega$.

five

IMPEDANCE
MEASUREMENTS

5-1 INTRODUCTION

Up to around 100 MHz the techniques used to make impedance measurements do not differ greatly from those used at audio frequencies. Impedance meters such as the HP 4115 RF vector impedance meter and the GR 1606-B radio-frequency bridge can be used, with due caution for stray coupling and loading effects. As the frequency is increased, the lines begin to become electrically significant in length, and the current and voltage, which must include phase information, become very difficult to measure with any accuracy. The measurement apparatus even becomes significant in size electrically and every circuit becomes a transmission line.

To deal with these unique situations, special measurement techniques have been developed. In this chapter we shall deal with probably the most common impedance measurement techniques presently employed. These are:

1. The slotted line.
2. The vector voltmeter.
3. The swept frequency method employing a frequency converter.

One of the simplest instruments used at high frequencies up into the GHz region is the slotted line. The time taken to make an impedance measurement can be quite large, but it has the advantage of being a very simple piece of hardware. As its name implies, the *slotted line* is an instrument made up of a section of line that has a longitudinal slot in its outer conductor. Its length should be at least a half-wavelength. A capacitive pickup probe is inserted a short distance into this slot and can be moved lengthwise. A typical slotted line and detection circuit is shown in Fig. 5-1.

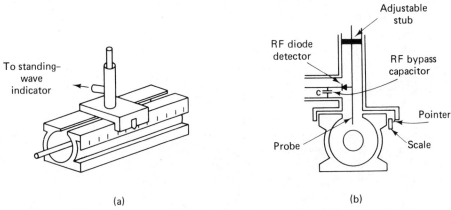

(a)   (b)

FIG. 5-1 Slotted line.

This probe, in parallel with the electric field (see Fig. 1-4), samples the electric field; and thus the voltage between the probe and its outer shield is proportional to the line voltage. Absolute voltages with such a device are not measured, only the ratio of the maximum voltage to the minimum voltage on the line. This ratio, as previously discussed, is the VSWR. The detection meters usually employed with the slotted lines permit this ratio to be read directly as a VSWR.

For proper VSWR readings, the probe must always project the same distance into the line as the probe is moved. Also, the probe should be kept to a minimal depth, so as not to alter the field in the line, resulting on an incorrect meter reading. This can be considered as a loading effect. A scale indicating distance is also located alongside the slotted line to keep note of the distance of the probe along the line. It is usually used to locate the null or node voltage points. These minimum voltage points are much sharper than the maximum voltage points and therefore can be more accurately located.

Often, the RF signal is AM-modulated (commonly at 1 kHz) to permit AC amplification by the detector circuitry. A tuned parallel circuit is located in the probe area to prevent loading of the main line. A crystal is usually used as the detector, and most VSWR meter indicators assume its response to be square-law in nature. For good accuracy, the crystal should be calibrated. This can be done by short-circuiting the line at some point beyond the slotted line and observing the meter reading as the probe is moved along the line. Assuming a lossless line, which is valid over short lengths at high frequencies, the standing wave in the slotted line will be sinusoidal. A comparison of the sine wave with the meter readings will provide the calibration curves.

The actual method of measuring an unknown impedance with the slotted line very closely follows the solution given in Example 4-4. The only difficulty still encountered is in obtaining the distance of the load to some point on the scale attached to the slotted line. This is determined by replacing the load by a short circuit and noting the minimum voltage points on the scale; this minima must be an integral number of half wavelengths from the load. Since impedances repeat every half wavelength, the load impedance can effectively be at any of these minima points. This whole procedure assumes a lossless line; if not, the attenuation must be taken into account.

The slotted-line technique can best be illustrated by an example.

EXAMPLE 5-1

With an unknown load attached to a 50-Ω slotted line, a VSWR of 5 is measured by a standing-wave indicator. The positions of the nulls are noted as indicated in Fig. 5-2(a). The distance between any null and the load remains unknown from these measurements only. We can, however, plot the VSWR circle of 5 on the Smith chart, knowing that any impedance on the line must lie somewhere on this locus.

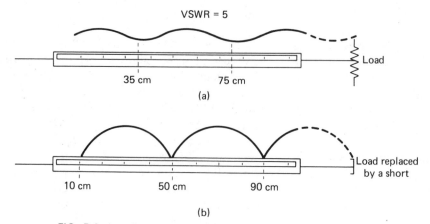

FIG. 5-2 Impedance measurements by the use of the slotted line.

The impedance at a voltage null is a minimum and can also be found on the Smith chart by locating the intersection of the VSWR circle with the zero reactance line at the minimum resistance point. This normalized impedance is seen to be 0.2 from Fig. 5-3.

To determine the distance from such a minimum to the load, the load is replaced by a short circuit [Fig. 5-2(b)]. By noting the new null positions and the fact that the dis-

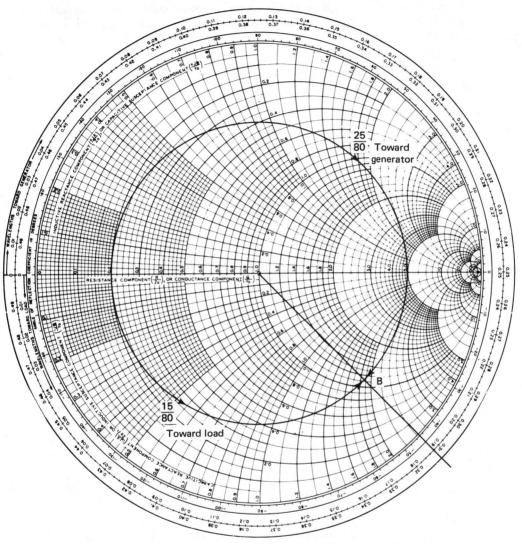

FIG. 5-3 Smith chart for Example 5-1.

tance between nulls represents $\lambda/2$, we can determine the distance to the short (or unknown, since they are, electrically speaking, at identical locations) from some point on the scale.

From Fig. 5-2(b), we see that the 90-cm position is $\lambda/2$ distant from the load (any integral number of wavelengths could be assumed). With this information we can now obtain the load impedance. For simplification we can note that the load, electrically speaking, could just as well be at the 90-cm or 50-cm or 10-cm locations, since the load impedance repeats itself every half a wavelength. The load impedance, therefore, can be found by rotating toward the load from a voltage minimum (impedance minimum) a distance of $(90 - 75)$ cm $= 15$ cm or $\frac{15}{80}\lambda = 0.1875\lambda$. Either direction of rotation can be used, and both are shown in Fig. 5-3. The value of the normalized load impedance is directly read as $1.13 - j1.89$. The load impedance for a Z_0 of $50\,\Omega$ is equal to $56.5 - j94.5\,\Omega$.

5-3 VECTOR VOLTMETER

Before describing the vector voltmeter technique of measuring complex impedances, the characteristics of the directional coupler should first be understood, since it is the heart of the measurement system.

Directional Coupler

The *directional coupler* is a passive device which has the facility to select either the incident wave or the reflected wave along a transmission line. The basic construction of a directional coupler consists of two major parts, the main arm and the auxiliary arm. Ideally, power flowing in the forward direction of the main arm is coupled to the output of the auxiliary arm, whereas power flowing in the reverse direction is not coupled to the auxiliary arm.

The principle of operation of a directional coupler is explained by use of Fig. 5-4. Although this two-slot device has a rather narrow bandwidth, it most

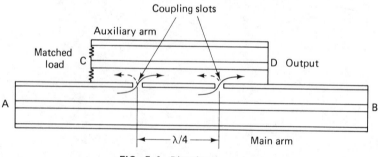

FIG. 5-4 Directional coupler.

easily portrays how the directional characteristics are obtained. In actual practice, a wideband directional coupler is employed which for coaxial lines consists of a system of precisely spaced bars to provide the coupling mechanism as shown in Fig. 5-5. It can also be made on strip transmission lines in a very similar fashion.

Referring to Fig. 5-4, waves moving from A through the two holes into the auxiliary line to point D arrive in phase and therefore reinforce each other. Practically no energy will appear in the reverse direction to point C, as the difference in path length is twice the hole spacing ($2 \times \lambda/4$), which causes a $180°$ phase difference between the waves, resulting in destructive interference.

For a reflected wave entering point B, there would be destructive interference at point D, but point C would see a reinforced wave proportional to the reflected

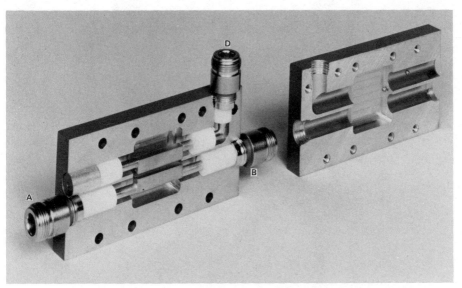

FIG. 5-5 Internal view of a Narda Microwave Corporation coaxial coupler.

signal in the main line. If part of the reflected wave is not to appear at point D, point C must be a good match.

The ratio, expressed in decibels of the forward power in the main arm to the sampled power in the auxiliary output, is called the *coupling factor* or *coupling coefficient* of the directional coupler. In practice, some reverse power in the main arm is coupled to the output of the auxiliary arm. The degree of discrimination in the auxiliary line of the power received in the auxiliary line for equal forward and reverse power is related to the "directivity" of the coupler. The ideal coupler will have an infinite value of directivity.

In equation form, these definitions in specifying directional couplers can be expressed as

$$\text{coupling factor} = 10 \log \frac{P_A}{P_{D1}} \qquad (5\text{-}1)$$

$$\text{directivity} = 10 \log \frac{P_{D1}}{P_{D2}} \qquad (5\text{-}2)$$

where P_{D1} = power in the auxiliary arm for forward power P_A in the main line
  P_{D2} = power in the auxiliary arm for equal reverse power in the main line

In making the measurements above, all terminals must be kept properly matched.

A typical directivity curve for a directional coupler is shown in Fig. 5-6. The finite directivity can result in measurement error, as the auxiliary arm does no longer solely contain a signal proportional to the reflected or the incident wave only. For most practical applications, a 30-dB directivity is adequate. The symbol often used for representing a directional coupler is shown in Fig. 5-7.

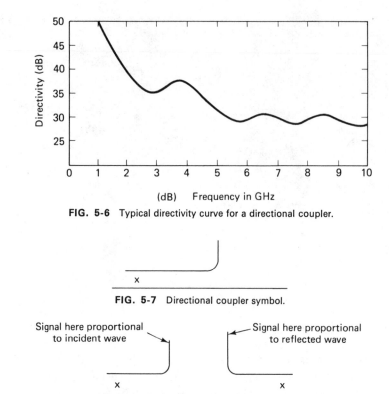

FIG. 5-6 Typical directivity curve for a directional coupler.

FIG. 5-7 Directional coupler symbol.

Signal here proportional to incident wave

Signal here proportional to reflected wave

FIG. 5-8 Dual directional coupler.

Dual directional couplers are also available that will sample both incident and reflected powers simultaneously (Fig. 5-8). In this case, arm C is not terminated as shown in Fig. 5-4, but is terminated in a detector.

Vector Voltmeter Impedance Measurement Technique

If a *vector voltmeter* (which measures both the voltage amplitude and phase difference between two points) is employed in conjunction with a dual directional coupler, a very simple impedance measuring system is available. The Hewlett-Packard 8405A vector voltmeter, for instance, measures the two voltages and the phase difference between its two input channels from 1 to 1000 MHz. By measuring the incident and reflected voltage at the unknown load terminals, we can calculate the load reflection coefficient ($\Gamma_R = E_R^-/E_R^+$) and from this determine the load impedance by plotting Γ_R on the Smith chart.

As it is impractical to measure incident and reflected voltages right at the load, we must make measurements some distance from it. Since impedances are repetitive every half-wavelength, we could put the measurement system $n\,\lambda/2$ back from the load (n being an integer). The disadvantage of this method is that upon a change in frequency, this physical distance must be altered.

A practical method of impedance measurement is shown in Fig. 5-9. It employs a line stretcher (adjustable 50 Ω air line) which is initially adjusted to equalize the signal path lengths from the generator to the sampling probes of the voltmeter $l_1 = l_2$. This calibration is made by temporarily replacing the load by a

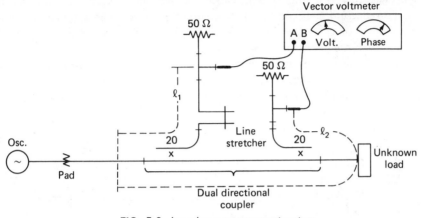

FIG. 5-9 Impedance measurement system.

short circuit ($\Gamma_R = -1$) and adjusting the line stretcher for a reference phase angle of 180° as read on the degrees phase indicator. This phase angle of 180° should remain constant as the frequency of the generator is varied. If not, the stretcher should be readjusted, as the signal path lengths are not equalized. With the load back in place, the incident and reflected voltages are again measured. The load reflection coefficient is then calculated and Z_R finally obtained. The latter can be done on the Smith chart or from the equation

$$\frac{Z_R}{Z_0} = \frac{1 + \Gamma_R}{1 - \Gamma_R} \tag{5-3}$$

As an example, let us assume after proper calibration, that the following measurements are made:

$$|E^+| = 50 \text{ mV}$$
$$|E^-| = 20 \text{ mV}$$

with the phase of E^- leading E^+ by 35°. Then

$$\Gamma_R = \frac{E^-}{E^+} = \frac{20 \times 10^{-3}}{50 \times 10^{-3}} \underline{/35°}$$

$$= 0.4 \underline{/35°}$$

Plotting this on the Smith chart, one obtains for Z_R/Z_0 the value

$$1.7 + j.91$$

If one assumes Z_0 to be 50 Ω,

$$Z_R = 50(1.7 + j.91) = 85 + j45.5 \text{ Ω}$$

5-4 SWEPT FREQUENCY TECHNIQUE

An improvement upon the two previously outlined point-by-point methods of impedance measurements is a swept frequency measurement system developed by Hewlett-Packard (HP). This system, called the *network analyzer*, permits one to obtain the reflection coefficient of the unknown impedance as a magnitude and phase quantity, or can plot the reflection coefficient directly on a Smith chart, depending upon the accessory used—all as a function of frequency.

Figure 5-10 shows a typical impedance measurement test setup. In this diagram the monopole antenna with the ground plane constitutes the unknown impedance. The reflection test unit shown in Fig. 5-10 contains a broad-band dual

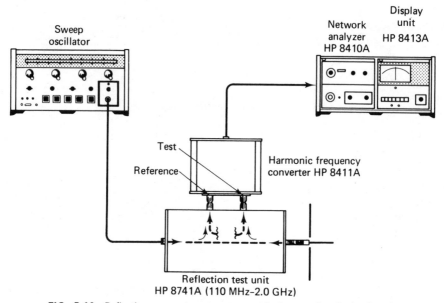

FIG. 5-10 Reflection measurement system employing Hewlett-Packard equipment.

directional coupler and a calibrated line stretcher. The adjustable line stretcher is for equalizing the electrical distance from the signal input to the incident and reflected outputs. In order to measure the load reflection coefficient with convenient low-frequency circuitry, the HP system converts the input RF signal by harmonic sampling. Since the frequency conversion employed is a linear process, the relative amplitudes and phases of the reference and test signals is preserved. A simplified block diagram of the system is shown in Fig. 5-11.

The internal phase-locked loop automatically tunes back and forth across the swept frequency band at a frequency separation of 20.278 MHz from the reference signal to maintain a constant IF frequency. After the second converter stage, the signal is kept at a constant 278 kHz for amplitude and phase determination. In the schematic shown, the reflection coefficient is read on the meter as a magnitude and phase quantity. The HP 8413A phase-gain indicator has also two dc outputs, which provide magnitude and phase information for an external recorder or dual-channel oscilloscope.

This network analyzer does not require a leveled signal source as the two matched AGC (automatic gain control) amplifiers provides equal changes of gain to both IF amplifiers while keeping the signal level of the reference channel constant. This allows the ratio of the test signal (reflected wave) to the reference signal (incident wave) to be measured directly on the meter indicator. The gain control, which provides accurate steps of attenuation, can be used to expand the output display resolution to ensure improved accuracy.

A polar display plug-in, HP 8414A (not shown), is also available which displays the plot of the reflection coefficient directly on a cathode ray tube. A Smith chart overlay may be used to permit ready conversion to impedance measurement. In all cases a continuous-wave or swept frequency mode can be employed. For a matched load, the CRT will display a dot at the center of the screen.

To calibrate the network analyzer, the load is temporarily replaced by a short, as is done also in the earlier techniques outlined. With the short in place ($\Gamma_R = -1$) the line stretcher in the reflection test unit is adjusted to read a Γ value also of -1 on the Smith chart over the desired frequency band. Initially, the scope trace is adjusted to trace a circular path around the periphery of the Smith chart, and finally reduced to a single point at the $0 + j0$ reactance of the Smith chart. The unknown load is then attached and the normalized impedance values can be read directly from the Smith chart ($Z_0 = 50\ \Omega$).

5-5 DETECTORS

As an amplitude-modulated wave is frequently used as a test signal in high-frequency measurements, as for instance in antenna pattern measurements and in the slotted line, it would be worthwhile to note a few of the detection schemes employed. We shall show that the output waveform from a square-law detector contains frequency components which are directly related to the power in the amplitude-modulated signal.

Assuming a modulation factor of m, where m is the ratio of the peak modulating voltage to the peak unmodulated carrier voltage (E_c), the amplitude-modulated waveform as depicted in Fig. 5-12 can be expressed by the equation

$$e = (E_c + mE_c \cos \omega_m t) \cos \omega_c t \qquad (5\text{-}4)$$

where ω_m is the radial modulating frequency and ω_c the radial unmodulated carrier frequency. Equation (5-4) can be expanded, using the trigonometric identity

$$\cos a \cos b = \tfrac{1}{2}[\cos (a + b) + \cos (a - b)] \qquad (5\text{-}5)$$

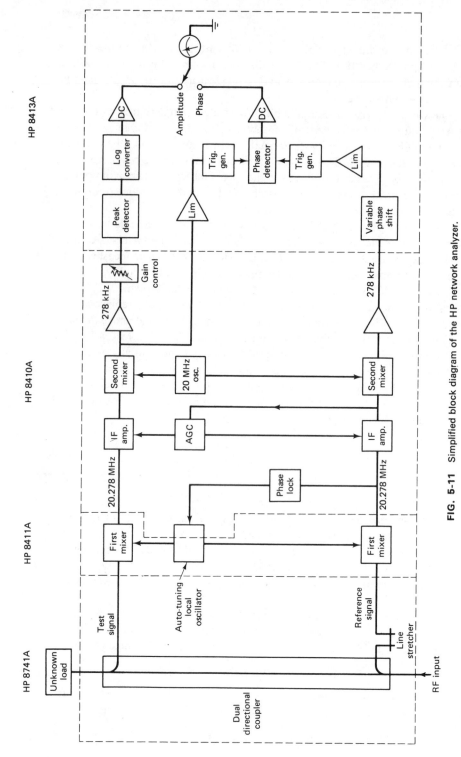

FIG. 5-11 Simplified block diagram of the HP network analyzer.

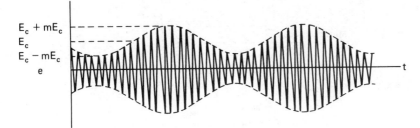

FIG. 5-12 AM test signal.

to give

$$e = E_c \cos \omega_c t + \frac{mE_c}{2} \cos (\omega_c - \omega_m)t + \frac{mE_c}{2} \cos (\omega_c + \omega_m)t \qquad (5\text{-}6)$$

The later equation indicates that three frequency components are present in the test signal:

1. The unmodulated carrier, $\omega_c/2\pi$.
2. The lower sideband, $(\omega_c - \omega_m)/2\pi$.
3. The upper sideband, $(\omega_c + \omega_m)/2\pi$.

The corresponding spectrum is shown in Fig. 5-13.

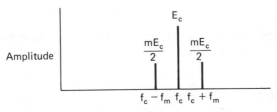

FIG. 5-13 Frequency spectrum of an AM signal.

In order to detect the AM signal, a crystal detector, as for instance shown in Fig. 5-1(b), is commonly employed. High-frequency crystals consist of a gold-plated tungsten wire (cat whisker), which makes a point contact with a semiconductor. The region of contact is filled with wax for rigidity and located in a ceramic housing as shown in Fig. 5-14. The crystal has a typical diode-response curve, as illustrated in Fig. 5-15, which when operated at very low levels results in square-law operation (i.e. the rectified output is proportional to the square of the RF input).

The voltage–current characteristic for the crystal diode around the origin can be expressed in terms of a MacLaurin's series. Thus,

$$i = a_0 + a_1 e + a_1 e^2 + \cdots + a_n e^n \qquad (5\text{-}7)$$

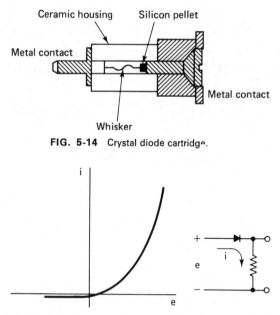

FIG. 5-14   Crystal diode cartridge.

FIG. 5-15   Voltage–current characteristic of a crystal diode.

where $a_0, a_1, \ldots, a_n$ are constants. Since the response curve has a current of 0 when the voltage is 0, a_1 must be 0. If the applied signal is sufficiently small as to operate in the square-law region, equation (5-7) can be simplified to

$$i = a_1 e + a_2 e^2 + \cdots \qquad (5\text{-}8)$$

by ignoring the higher-order terms. Thus, prior to filtering, the output from the diode detector (e'_0) shown in Fig. 5-16 is equal to

$$e'_0 = iR = a_1 Re + a_2 Re^2 \qquad (5\text{-}9)$$

To obtain the frequency components at the detector output with an AM input signal, equation (5-6) must be substituted into equation (5-9) and the appropriate algebra performed. In addition to identity (5-5), the $\cos^2 a$ identity must be used:

$$\cos^2 a = \frac{1 + \cos 2a}{2} \qquad (5\text{-}10)$$

Thus,

$$e'_0 = a_1 R \left[E_c \cos \omega_c t + \frac{mE_c}{2} \cos (\omega_c - \omega_m)t + \frac{mE_c}{2} \cos (\omega_c + \omega_m)t \right]$$

$$+ a_2 R \left[E_c \cos \omega_c t + \frac{mE_c}{2} \cos (\omega_c - \omega_m)t + \frac{mE_c}{2} \cos (\omega_c + \omega_m)t \right]^2$$

$$= a_1 R E_c \left[\cos \omega_c t + \frac{m}{2} \cos (\omega_c - \omega_m)t + \frac{m}{2} \cos (\omega_c + \omega_m)t \right.$$

$$ \{f_c\} \qquad\qquad \{f_c - f_m\} \qquad\qquad \{f_c + f_m\}$$

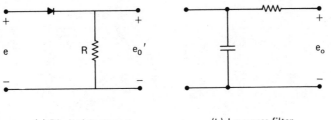

(a) Diode detector   (b) Low pass filter

FIG. 5-16 Crystal detector circuit.

$$+ a_2 R E_c^2 \Big[\cos^2 \omega_c t + \Big(\frac{m}{2}\Big)^2 \cos^2 (\omega_c - \omega_m) t + \Big(\frac{m}{2}\Big)^2 \cos^2 (\omega_c + \omega_m) t$$
$$\{dc, 2f_c\} \qquad \{dc, 2(f_c - f_m)\} \qquad \{dc, 2(f_c + f_m)\}$$

$$+ m \cos \omega_c t \cos (\omega_c - \omega_m) t + m \cos \omega_c t \cos (\omega_c + \omega_m) t$$
$$\{f_m, 2f_c - f_m\} \qquad\qquad \{f_m, 2f_c + f_m\}$$

$$+ \frac{m^2}{2} \cos (\omega_c - \omega_m) t \cos (\omega_c + \omega_m) t \Big] \qquad\qquad (5\text{-}11)$$
$$\{2f_m, 2f_c\}$$

The frequencies contributed by each term are noted in { } under the corresponding terms in the last equation. By employing identities (5-5) and (5-10), the amplitudes of each frequency component are noted in Table 5-1. The corresponding typical spectrum plot appears as that shown in Fig. 5-17. For simplification of amplifier circuitry, one of the lower frequencies is chosen for detection purposes. The dc component is usually not used when amplification is desired due to the drift problem in dc amplifiers. It is common practice to select the modulating frequency f_m.

The low-pass filter shown in Fig. 5-16(b) removes all the high-frequency components, leaving only the dc, f_m, and $2f_m$ frequencies. Any one of these components

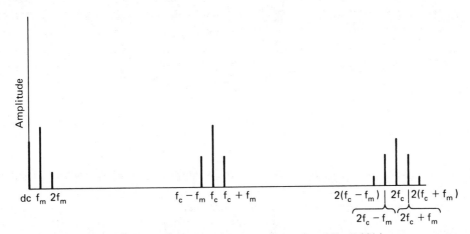

FIG. 5-17 Frequency spectrum of e_0' from diode detector of Fig. 5-16(a).

TABLE 5-1 Amplitudes of Various Frequency Components

Frequency component	Peak amplitude
dc	$a_2 RE_c^2\left(\dfrac{1}{2}+\dfrac{m^2}{4}\right)$
f_m	$a_2 RE_c^2\, m$
$2f_m$	$a_2 RE_c^2\,\dfrac{m^2}{4}$
$f_c - f_m$	$a_1 RE_c\,\dfrac{m}{2}$
f_c	$a_1 RE_c$
$f_c + f_m$	$a_1 RE_c\dfrac{m}{2}$
$2f_c - f_m$	$a_2 RE_c^2\,\dfrac{m}{2}$
$2f_c$	$a_2 RE_c^2\left(\dfrac{1}{2}+\dfrac{m^2}{4}\right)$
$2f_c + f_m$	$a_2 RE_c^2\,\dfrac{m}{2}$
$2(f_c - f_m)$	$a_2 RE_c^2\,\dfrac{m^2}{8}$
$2(f_c + f_m)$	$a_2 RE_c^2\,\dfrac{m^2}{8}$

has an amplitude that is proportional to the square of the carrier voltage or the power in the test signal. A typical example is a VSWR meter which is sharply tuned (25-Hz bandwidth) at 1 kHz. It amplifies only the f_m component and effectively takes the square root of the amplitude to obtain a reading directly related to the voltage of the test signal. The VSWR meter can read VSWR directly by employing it with a slotted line. The probe is set in the slotted line to a maximum in the standing-wave pattern and the amplifier gain adjusted for full-scale reading. The slotted line probe is then moved to a minimum and the VSWR read directly. For large VSWR measurements, the dB difference between the maximum and minimum voltage points is noted and converted to the equivalent VSWR by the relation

$$\text{VSWR} = \text{antilog}\left(\frac{\text{dB difference}}{20}\right) \tag{5-12}$$

The dc component is chosen in instances where the signal is sufficiently strong so as not to require any amplification.

As can be seen from the many components present from the crystal output, crystal diodes can also be used as microwave mixers. The incoming microwave signal mixes with a local oscillator signal in the crystal to produce the sum and difference frequencies, as well as many other frequencies. A bandpass filter is then used to retain the desired frequency component.

Square-wave or on–off modulation is also frequently employed to reduce the spurious FM that is produced by the signal source when employing lower-level modulation. The mathematics becomes slightly more involved and will not be attempted, but the results are similar to those just derived. More components are present, but the dc term and f_m terms again relate to the square of the carrier voltage.

Measurement of Power

To measure microwave power in the milliwatt region, a bolometer or temperature-sensitive resistive element is frequently employed. Larger powers can be monitored by adding a directional coupler. Bolometers are nominally square-law devices; that is, they produce output currents proportional to the incident microwave power. There are two main types of bolometers:

1. *Barretter*—having positive temperature coefficient of resistance.
2. *Thermistor*—having negative temperature coefficient of resistance.

The characteristics of a typical thermistor and a barretter are shown in Fig. 5-18. The figure shows that the thermistor is the more sensitive of the two devices. The resulting large mismatch when a thermistor is overloaded causes it to have excellent burnout protection characteristics. A barretter, on the other hand, is more susceptible to burnout, but is more reproducible in both sensitivity and impedance, and is less sluggish (has a small thermal time constant) than thermistors.

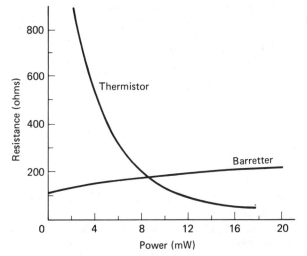

FIG. 5-18 Typical bolometer characteristics at an ambient temperature of 20°C.

The barretter consists of a thin platinum wire or a thin metallic film on glass. It is nominally biased to operate at a resistance of about 200 Ω. When measuring power-level ratios, as in the case of antenna radiation patterns, the RF signal is modulated and the barretter is used as a detector tuned to the modulating frequency. The barretter must be able to follow the modulation envelope so as to be able to obtain a usable audio-frequency output. This limits the modulating frequency to a maximum of 2 kHz. A typical barretter detector circuit is shown in Fig. 5-19. In the direct-reading detection circuit, the accuracy of measurement depends chiefly on the deviation of the barretter from the square law.

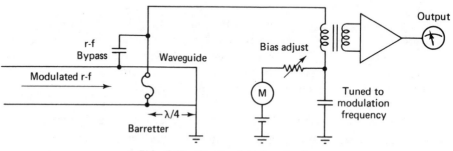

FIG. 5-19 Barretter detection circuit.

A bead thermistor element consists of a glass-enclosed semiconducting material supported by two taut wires, as shown in Fig. 5-20. The coating is to protect it from oxidation and make it stable. Thermistors are also usually biased at about 200 Ω. To avoid mismatch problems, the bolometer is usually placed in one arm of a bridge circuit and biased so as to maintain a constant resistance of, say,

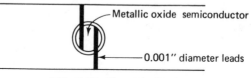

FIG. 5-20 Thermistor bead.

200 Ω. Initially, the bridge is balanced with dc power and audio-frequency power (e.g., 10.8 kHz) with no RF field present. Application of RF power causes the bolometer to heat up and change resistance and thus unbalance the bridge. The feedback mechanism in the bridge then reduces the output from the audio oscillator to cause the bridge to fall back in balance. The reduction in audio power from the oscillator indicates the amount of RF field applied to the bolometer. A typical simplified schematic is shown in Fig. 5-21.

The thermistor in the dc power section compensates for any changes in ambient temperature. If the temperature increases, the resistance of the thermistor

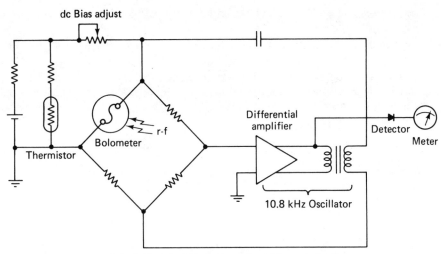

FIG. 5-21 Simplified schematic of RF power bridge.

decreases, causing less dc current to flow into the bridge, thus maintaining bridge balance.

PROBLEMS

5-1. With the load in place on a slotted line, the value of VSWR is 4.0, and the minima occur at 15, 25, 35 cm, and so on, on the scale, increasing toward the load. With the load temporarily replaced by a short circuit, the minima are found at 12, 22, 32 cm, and so on. Find the load impedance if Z_0 is 50 Ω. Also determine the frequency. (*Note:* $v_p = c$.)

5-2. The slotted line technique is used to measure an unknown impedance. Its scale increases toward the load. The following observations were made:

    With load: VSWR $= 2$
            Voltage minima at 40, 60, 80 cm, ...
    With load replaced by short: VSWR $= \infty$
            Voltage minima at 45, 65 cm . . . .

Find the normalized load impedance.

5-3. (a) 100 watts of RF power is applied through a directional coupler to a matched load. If the power output from the coupled secondary port is 1/2 W, what is the coupling factor (in dB) of the directional coupler?

(b) By reversing the directional coupler, 100 W flows through the coupler in the opposite direction. If the output power in the same secondary port is now 20 $\times 10^{-6}$ W, what is the directivity of the coupler (in dB)?

5-4. If the line stretcher in Fig. 5-9 is not available, explain by the way of an example how you would go about obtaining an unknown load impedance using the same equipment as shown.

5-5. Explain the proper calibration procedure of the measurement system shown in Fig. 5-11.

5-6. A directional coupler shows an incident power of 10 W. If the VSWR on the main line is 3.0, what power is being absorbed by the load? [*Note:* The directional coupler monitors the incident power $|E^+|^2/Z_0$ and the power absorbed by the load will be $(|E^+|^2 - |E^-|^2)/Z_0$.]

six

IMPEDANCE MATCHING

6-1 INTRODUCTION

It has been noted that when a transmission line is not properly terminated, a reflected wave exists on the line. This can cause difficulties, depending upon the application for which the line is being used. To eliminate these reflections, impedance matching is frequently employed. Some of the benefits of matching are:

1. The transmission line transmits a given power with a smaller peak voltage, and consequently there is less chance of flashover at large values of power.

2. The input impedance remains at the value Z_0 in a "flat" line when the frequency changes (i.e., there is no frequency modulation distortion).

3. The nonresonant lines do not tend to "frequency-pull" the generator from its nominal value. A resonant line has a variable reactive component with frequency shift which reflects into the generator's output circuitry.

In addition to the features above, there is also a more efficient transfer of power down the transmission line.

When maximizing power transfer from a generator to a load, however, an additional consideration must be taken into account. As proven in most introductory circuit-theory textbooks, maximum power is transfered to a load such as that shown in Fig. 6-1 when the load impedance is the complex conjugate of the generator impedance. That is, when

$$R_R = R_G \qquad (6\text{-}1)$$

$$X_R = -X_G \qquad (6\text{-}2)$$

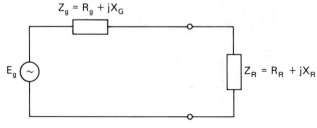

FIG. 6-1   Generator with its load.

Thus, to obtain both maximum power transfer from the generator to the load and no reflection, suitable matching sections or impedance transformers must be located at the generator and the load points in a transmission system. Figure 6-2 shows a properly matched low-loss transmission system. For such a case, Z_0 is real and the impedances at every point are conjugates when looking in opposite directions.

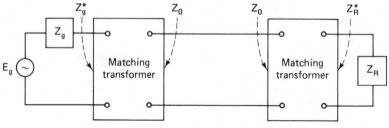

FIG. 6-2   Completely matched transmission line.

In practice, commercial RF signal generators and power amplifiers, and so on, have an output impedance that corresponds closely to the characteristic impedance of commercially manufactured transmission lines. What often remains to be done is to match some mismatched load to the system. At times a load impedance can be adjusted to match the characteristic impedance of the line, but frequently this is not possible. A matching device must then be inserted near the load to transform the load impedance to the characteristic impedance of the line.

At low frequencies, the iron-cored transformer is chiefly used as the impedance matching device. At higher frequencies, up to a few hundred megahertz or so, the air-cored transformer and the ferrite transformer are employed (see, for instance, the remarks made on the balun in Section 3-7). If a balun is not required, four terminal networks consisting of reactive elements in a Π or T configuration are frequently used. As the frequency is increased, however, owing to stray coupling and distributed effects, these devices become unpredictable and short sections of transmission lines are used. In the remaining sections the following matching devices will be considered.

1. The quarter-wave transformer.
2. The single stub tuner.
3. The double stub tuner.
4. The exponential taper or broad-band transformer.

6-2 QUARTER-WAVE TRANSFORMER

The *quarter-wave transformer* consists of a transmission line $\lambda/4$ long, as shown in Fig. 6-3. From equation (3-54) we can find the input impedance of a lossless quarter-wave transmission line terminated in an impedance Z_R. Thus, since $\alpha = 0$,

$$Z_s = Z_0 \frac{0 + jZ_0}{0 + jZ_R} = \frac{Z_0^2}{Z_R} \qquad (6\text{-}3)$$

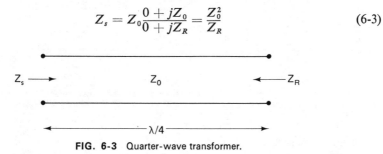

FIG. 6-3 Quarter-wave transformer.

where 0 has been substituted for $\cos \beta l$ and 1 has been substituted for $\sin \beta l$. (*Note:* $\beta l = 2\pi/\lambda \times \lambda/4 = \pi/2$.)

If the load impedance $(Z_R = R_R)$ and the characteristic impedance of the quarter-wave line are both real, the input impedance Z_s is also real and equal to

$$Z_s = \frac{Z_0^2}{R_R} \qquad (6\text{-}4)$$

A mismatched load can be properly matched to a transmission line by inserting prior to the load a quarter-wave transformer having a suitably selected characteristic impedance. The required characteristic impedance of this quarter-wave section can be obtained from equation (6-3) and is the geometric mean of the termination and desired sending-end impedance.

$$Z_0 = \sqrt{Z_R Z_s} \qquad (6\text{-}5)$$

As it is difficult to construct a transmission line with the complex characteristic impedance as normally obtained by equation (6-5), both Z_R and Z_s should be made real. The desired sending-end impedance is usually real, as it is equal to the characteristic impedance of the main line. A little later we shall see how Z_R is made real. Thus, for real impedances,

$$Z_0 = \sqrt{R_R R_s} \qquad (6\text{-}6)$$

As an example, a 100-Ω load can be matched to a 50-Ω transmission line by inserting a $\lambda/4$ section of line between the load and the main line. This quarter-wave transformer must have a characteristic impedance of $\sqrt{50 \times 100} \approx 71$ Ω.

It is not too difficult to alter the characteristic impedance of an open-wire line, as this can be done by changing the conductor diameter and spacings. With more difficulty, the impedance of a coaxial line can be changed by altering the diameter of the inner or outer conductor. This can be done by appropriate machining or inserting a sleeve around the inner or outer conductor as shown in Fig. 6-4. A slug of dielectric material with a suitable dielectric constant could also be inserted. It should be noted that the wavelength in such a region is different than in the rest of the line.

FIG. 6-4 Quarter-wave coaxial transformers.

A complex load can be matched to a line by transforming the load to a real impedance by leaving a suitable extension of the main line to the load before inserting the quarter-wave transformer. An illustration using the Smith chart is shown in Fig. 6-5. The chart is entered at Z_R/Z_0' and the impedance locus is followed toward the generator to a point where the input impedance is purely resistive. The distance traveled is the length l. Either of the two intersections may be used, but one is generally more practical than the other. The characteristic impedance of the quarter-wave section is either

$$Z_0 = \sqrt{R_1 Z_0'} \quad \text{or} \quad \sqrt{R_2 Z_0'}$$

The chief disadvantage of the quarter-wave transformer is that it is a narrow-band device. Frequency sensitivity can be reduced by using a severel section transformer. For a two-section transformer, as shown in Fig. 6-6, the best results are obtained when

$$\left(\frac{Z_0^{11}}{Z_0^{1}}\right)^2 = \frac{Z_0^{1}}{Z_0} = \left(\frac{Z_0}{R}\right)^2 \qquad (6\text{-}7)$$

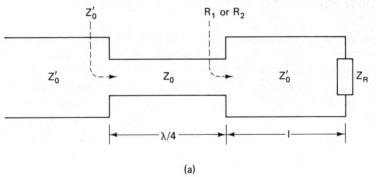

(a)

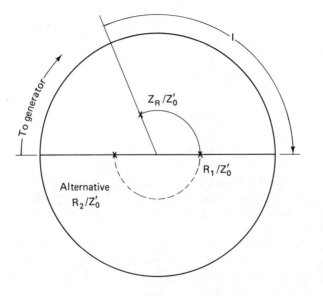

(b)

FIG. 6-5 Application of $\lambda/4$ transformer.

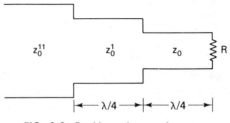

FIG. 6-6 Double section transformer.

111

6-3 SINGLE STUB TUNER

The *single stub tuner* illustrated in Fig. 6-7 consists of an open- or short-circuited section of transmission line shunted across the main line some distance l_1 from the load. Although a series stub would be theoretically feasible, it is difficult, if not impossible, to insert in a coaxial line; and hence it will not be pursued here. Also, short-circuited stubs are preferred to open-circuited stubs, as a true open is virtually impossible to achieve, owing to radiation from the open end. Both the distance from the load to the stub, l_1, and the stub length l_2, must be variable.

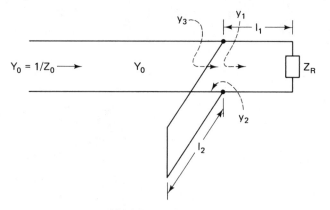

FIG. 6-7 Single stub tuner.

Computations for the single stub tuner are easily made with the use of the Smith chart. Since a parallel connection between the stub and the line are involved, admittances are more convenient to use than impedances, as admittances in parallel can be directly added.

As has been explained in Section 3-7, a lossless shorted transmission line contributes only a reactance or a susceptance component. Referring to Fig. 6-7, for a proper match, the normalized admittance $y_3 = Y_3/Y_0$ must be equal to unity. Since the stub adds susceptance only, the normalized admittance y_1 must differ from y_3 by the normalized susceptance only. (Lowercase letters refer to normalized values.) The length l_1 is selected so that the normalized admittance y_1 has a real part equal to unity. That is, looking toward the load at the stub junction, the normalized admittance is forced to be

$$y_1 = \frac{Y_1}{Y_0} = 1 + jb \qquad (6\text{-}8)$$

where b is an undesired susceptance. The length of the shorted stub (Fig. 6-8) is then adjusted so that the normalized susceptance component of y_1, that is, jb, is canceled out.

$$y_2 = \frac{Y_2}{Y_0} = -jb \qquad (6\text{-}9)$$

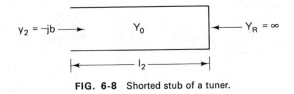

FIG. 6-8 Shorted stub of a tuner.

EXAMPLE 6-1

Consider the problem of matching a $100 + j50\,\Omega$ load to a $50\,\Omega$ line, using a single stub as shown in Fig. 6-9.

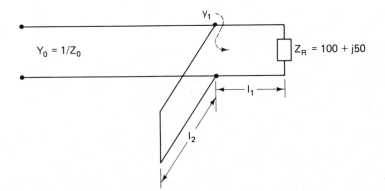

FIG. 6-9 Example problem of a single stub match.

Solution: The normalized impedance $(100 + j50)/50 = 2 + j1$ is plotted on the Smith chart (see Fig. 6-10) and the corresponding normalized admittance is obtained from the point opposite on the chart ($y_R = Y_R/Y_0 = 0.4 - j.2$).

To be able to obtain a match, y_1 must take the form $1 \pm jb$. A constant VSWR or $|\Gamma|$ locus is followed from the load toward the generator until it intersects the unity conductance circle. This can always occur at two points, and in this case the first intersect, $y_1 = 1 + j1$, will be chosen. This point occurs $(0.162 + 0.037)\lambda = 0.199\lambda$ from the load and is the location where the stub is placed.

The normalized susceptance $+j1$ must be canceled by the stub (Fig. 6-11). Therefore, the normalized susceptance looking into the stub must be $-j1$. l_2 is obtained by locating $Y_R/Y_0 = \infty$ on the Smith chart and rotating toward the generator on a VSWR $= \infty$ circle until a normalized susceptance of $-j1$ is reached. From the Smith chart this is seen to be 0.125λ. Thus,

$$l_1 = 0.199\lambda$$

$$l_2 = 0.125\lambda$$

Single stub impedance matching requires that the stub be located at a definite point on the line. Even after proper calculations of the stub's position and with careful location of the stub in the field it is usually necessary to make minor adjustments to obtain a good match. It is generally not feasible to make any small adjustments

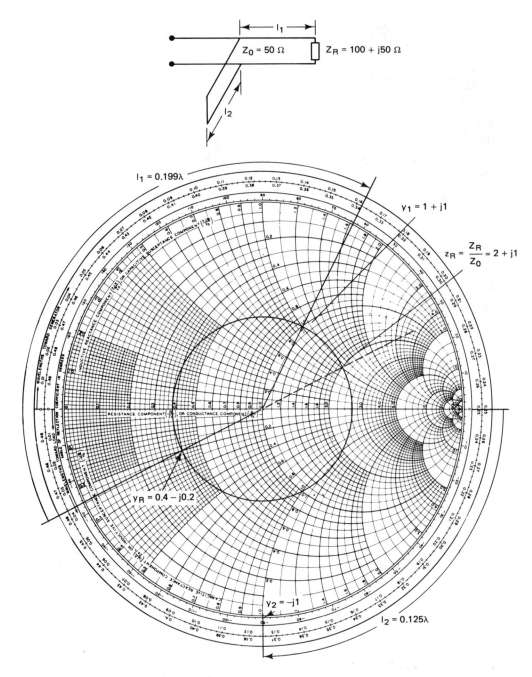

FIG. 6-10 Smith chart for Example 6-1.

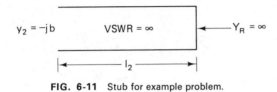

FIG. 6-11   Stub for example problem.

of the stub position on the line once it is inserted, and so another method is used in which two stubs at arbitrary positions are used. Such a tuner is discussed in the next section.

6-4   DOUBLE STUB TUNER

The *double stub tuner* is a matching network consisting of two adjustable stubs fixed in position on the transmission line. Often in practice, small mica capacitors, which can be considered as a short length of open-circuited transmission line, are connected to the ends of a predetermined length of stub line to allow for some small "tweaking" in the field. Although the spacing of the stubs is not critical, an odd number of eighth-wavelengths will match a wide range of impedances. In our examples we shall use a stub separation of $\lambda/8$ as shown in Fig. 6-12. For a proper

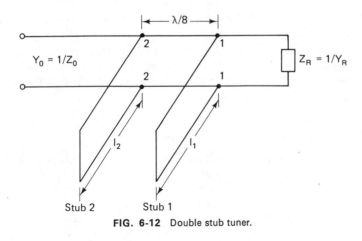

FIG. 6-12   Double stub tuner.

match, the input normalized admittance to the left of junction 2–2 should be unity. Since stub 2 adds susceptance only to the line, the normalized admittance to the right of junction 2–2 must be of the form $1 + jb$. Hence, the admittance to the right of the 2–2 junction must appear on the normalized conductance locus of 1, as shown by the dashed circle A on the Smith chart of Fig. 6-13.

The transformer formed by the one-eighth wavelength of line between 2–2 and 1–1 will transform all the admittances that lie on circle A to points on circle B, which is displaced one-eighth wavelength toward the load from circle A. Therefore,

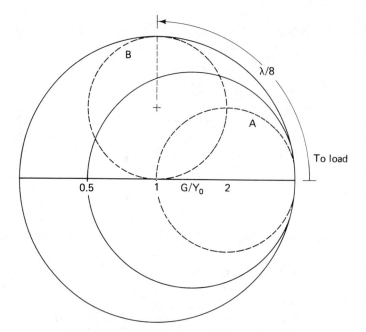

FIG. 6-13   Loci of importance for the double stub.

if stub 1 succeeds in transforming the input admittance of the line and load to the right of junction 1–1 into an admittance that will plot on the circle B locus, the eighth wavelength line will further transform the admittance into a value just to the right of junction 2–2, which will plot on the locus circle A and will have a normalized admittance of $Y/Y_0 = 1 + jb$. Stub 2 must then provide for cancellation of the susceptive component, $+jb$.

EXAMPLE 6-2

The normalized admittance of the load on a line is $0.3 - j2.0$. This is to be matched to the line by a double stub spaced $\lambda/8$ apart, with nearest stub 0.1λ from the load.

Solution: Since the stub spacing is $\lambda/8$, circle A (see Fig. 6-17) is rotated toward the load $\lambda/8$ to form circle B. The admittance just to the left of stub 1 must lie on locus B. The admittance seen to the right of stub 1 is found by entering the Smith chart at $0.3 - j2.0$ and rotating 0.1λ toward the generator on a constant VSWR circle. The normalized admittance to the right of 1–1 is therefore $0.08 - j0.53$ (point 1 on the Smith chart.) Stub 1 adds a susceptance to the line which must change the latter admittance to a value that lies on circle B. Stub 1 cannot alter the conductance, so following a constant conductance circle from point 1 locates point 0 on circle B, where

$$y_1 = \frac{Y_1}{Y_0} = 0.08 + j0.6$$

Notice that the conductance has not changed.

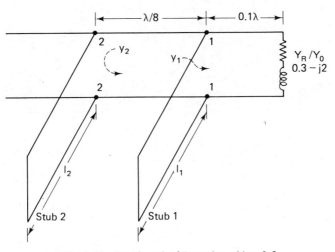

FIG. 6-14 Double stub of Example problem 6-2.

Thus, stub 1 must contribute a normalized susceptance of

$$(0.08 + j0.6) - (0.08 - j0.53) = j1.13$$

(see Fig. 6-15). From the Smith chart, l_1 is found to be 0.385λ (rotating along the infinite VSWR circle toward the generator from the infinite admittance point until an admittance of $j1.13$ is found).

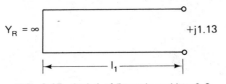

FIG. 6-15 Stub 1 of Example problem 6-2.

The change in admittance caused by the section of line between junctions 1–1 and 2–2 can be obtained by following a constant VSWR circle, which in this case is 17, a distance of $\lambda/8$ toward the generator, resulting in a value that must lie on circle A. The admittance at 2–2 on the line, without stub 2 connected, can be read from the chart at point 2 on the locus circle A, giving

$$y_2 = \frac{Y_2}{Y_0} = 1 + j3.9$$

Stub 2 should then be adjusted to provide a normalized inductive susceptance of $-j3.9$ to cancel out the remaining susceptance of $+j3.9$, after which the admittance to the left of

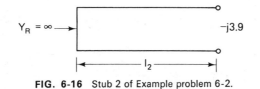

FIG. 6-16 Stub 2 of Example problem 6-2.

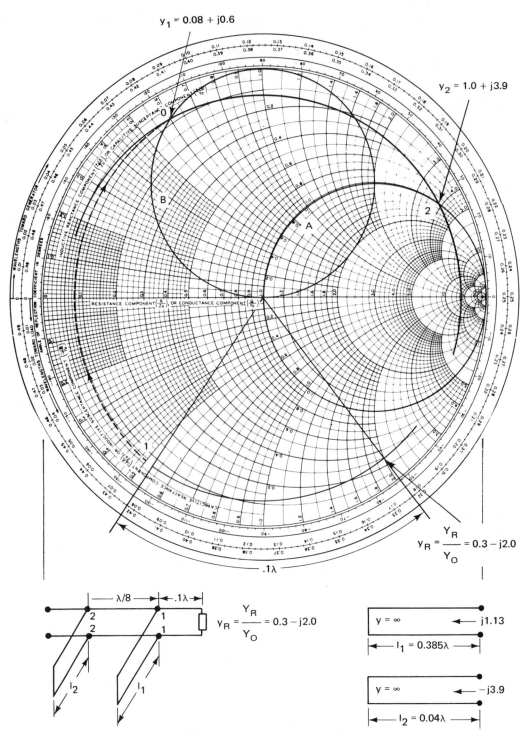

FIG. 6-17 Smith chart for Example 6-2.

2–2 will be $1 + j0$. From the Smith chart, l_2 is found to be 0.04λ (rotating along the infinite VSWR circle toward the generator from the infinite admittance point until an admittance of $-j3.9$ is achieved).

The double stub tuner suffers from the disadvantage that not all loads can be matched. This can be overcome by making this distance to the load adjustable or by adding a third stub. The double stub tuner is also inherently a narrow-band device. In order to achieve broadbanding, the tapered line discussed in the next section can be employed.

6-5 EXPONENTIAL TAPER

The *exponentially tapered line* is a line in which the characteristic impedance varies exponentially along its length. If the taper per wavelength is small, such a line can provide impedance transformation that is relatively insensitive to frequency. The exponentially tapered line as shown in Fig. 6-18 may have its inductance and

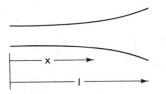

FIG. 6-18 Exponentially tapered line.

capacitance per unit length at any point x expressed as

$$L(x) = L(0)e^{\delta x} \tag{6-10}$$

$$C(x) = C(0)e^{-\delta x} \tag{6-11}$$

where $L(0)$ and $C(0)$ are the values at the narrow end and δ indicates the rate of taper. Refering back to equation (3-18), the characteristic impedance can therefore be expressed as

$$Z_0(x) = \sqrt{\frac{L}{C}} = \sqrt{\frac{L(0)}{C(0)}}\, e^{\delta x}$$

$$= Z_0(0)e^{\delta x} \tag{6-12}$$

For a line of length l, a transformation ratio, T, can be defined as

$$T = \frac{Z_0(l)}{Z_0(0)} = e^{\delta l} \tag{6-13}$$

where $Z_0(l)$ = characteristic impedance at the high-impedance end
$Z_0(0)$ = characteristic impedance at the low-impedance end

Associated with such a taper is its behavior, like that of a high-pass filter. It freely passes energy above a certain critical frequency called the *cutoff frequency*, but attenuates the signals below this frequency.[1] The cutoff frequency is given by

$$f_{cutoff} = \frac{\delta c}{4\pi\sqrt{\epsilon_r}}$$ (6-14)

where c = velocity of light

ϵ_r = relative dielectric constant of the transmission line

The cutoff frequency should be three or four times lower in frequency than the frequency of operation.

Combining equations (6-13) and (6-14) yields the physical length of the line:

$$l = \frac{\ln T}{\delta} = \frac{c \ln T}{4\pi\sqrt{\epsilon_r} \, f_{cutoff}}$$ (6-15)

An illustration of employing an exponential taper on a microstrip line is shown in Fig. 6-19. In this particular case, the impedance is transformed from $1.2 + j0 \, \Omega$ to $50 + j0 \, \Omega$ over a frequency range of 700 to 1400 MHz by employing 25-mil-thick alumina ($\epsilon_r = 9.9$).

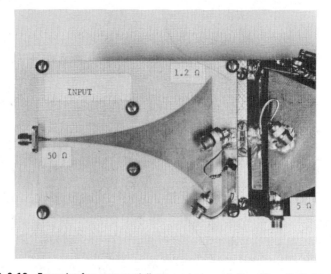

FIG. 6-19 Example of an exponentially tapered microstrip line. (From R. P. Arnold and W. L. Bailey, "Match Impedances with Tapered Lines," *Electronics Design,* Vol. 22, No. 12, June 7, 1974, 136–139.)

---

[1] For more detailed analysis of the tapered line, see W. C. Johnson, *Transmission Lines and Networks* (New York: McGraw-Hill Book Company, 1950).

PROBLEMS

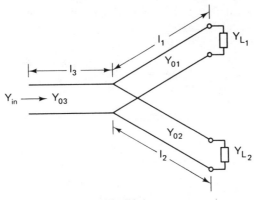

FIG. P6-1

6-1. Find Y_{in} for the multiple-section line shown, assuming the following values:

$$l_1 = 0.10\lambda \qquad l_2 = 0.20\lambda \qquad l_3 = 0.08\lambda$$

$$Y_{01} = 0.01 \text{ S} \qquad Y_{02} = 0.01 \text{ S} \qquad Y_{03} = 0.02 \text{ S}$$

$$Y_{L_1} = 0.002 + j.003 \text{ S} \qquad Y_{L_2} = 0.003 - j.003 \text{ S}$$

(*Hint:* It is convenient to solve this type of problem in terms of admittances.)

6-2. Find the characteristic impedance Z_0' and position of a quarter-wave transformer required to match the load $Z_L/Z_0 = 0.3 + j.3$ to a line having $Z_0 = 50 \, \Omega$ if the transformer is placed as close to the load as possible.

6-3. When power is to be applied to more than two loads, multiple branching at some point is not feasible, since (1) it is difficult to mechanically construct such a branch, and (2) the characteristic impedance of the branching lines becomes impractically high if the main line is to be matched. One method of multiple load power division is shown below. This distance between each branch point is $\lambda/4$.

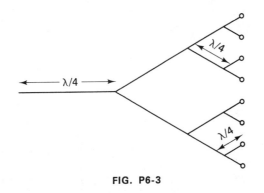

FIG. P6-3

121

(a) If each load is 100 Ω and the characteristic impedances of each section of line is 70 Ω, what is the input impedance to the network?

(b) What is the phase relationships of the signal to each load?

6-4. Determine the length of a 50 Ω short-circuited line required to produce a
(a) Susceptance of $+j.001$ S.
(b) Susceptance of $-j.005$ S.

6-5. Find the length and position of a stub in wavelengths used to match a normalized load of $0.3 + j0.3$ to a transmission line. The stub should be located as close to the load as possible.

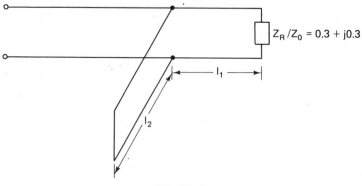

FIG. P6-5

6-6. (a) Find the length (l_2/λ) and position (l_1/λ) of a 50 Ω shorted stub $(l_2/\lambda < 1/4)$ which will match the load.

(b) Find the length and position of a 50 Ω open stub that will match the load. Keep the stub length under $\lambda/4$.

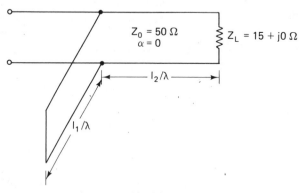

FIG. P6-6

6-7. For the double stub tuner shown, find the lengths of L_1 and L_2 in terms of wavelengths necessary to match the 50 Ω line to load impedance of $100 + j50$ Ω.

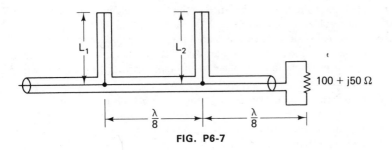

FIG. P6-7

6-8. An impedance $Z = 41 + j15\ \Omega$ is connected as in the diagram shown, to a coaxial line of $Z_0 = 100\ \Omega$. Find the lengths of l_1 and l_2 (in cm) to match the line to the load, by use of the Smith chart. (*Note:* 20 cm $= \frac{3}{8}\ \lambda$.)

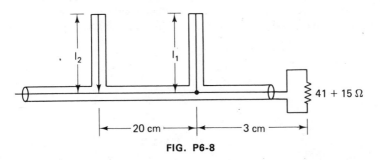

FIG. P6-8

6-9. Two quarter-wavelength transformers in tandem are to be designed to connect a 25 Ω resistive load to a 100 Ω line. Determine the characteristic impedance for each section.

6-10.

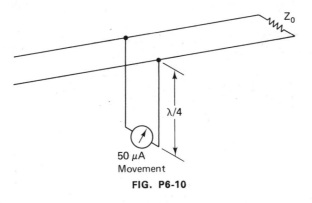

FIG. P6-10

A low-loss transmission line is matched with a characteristic impedance of 300 Ω. An RF voltmeter is attached as shown, which consists of a 300 Ω line one quarter-wavelength long. The meter has a resistance of 10 Ω with a full-scale deflection of 50 μA.

(a) Determine the input impedance of the RF voltmeter and calculate the VSWR that results on the main line.

(b) What main-line voltage is required for full-scale deflection?

(c) If the meter movement is square-law (i.e., the deflection is proportional to the square of the voltage applied to the meter), what line voltage will produce half-scale deflection?

6-11. A tapered microstrip transmission line (see Fig. 6-19) is to be used as an impedance transformer between a 1.2 Ω load and a 50 Ω coaxial cable. Using 25-mil (0.025 in.) alumina as the substrate ($\epsilon_r = 9.9$), determine the length and the end widths of the microstrip conductor if the cutoff frequency is to be one-fifth of the 1.5-GHz operating·frequency.

6-12. A tapered microstrip transmission line is to be used as a broad-band impedance transformer from the output of a transistor amplifier (5 Ω) to a 50 Ω coaxial cable. Using Teflon–Fiberglas having an ϵ_r of 2.55 and a thickness of 10 mm as the dielectric, determine the length and the end widths of the microstrip conductor required if the cutoff frequency is one-fifth of the 1.5-GHz operating frequency.

seven

PLANE WAVES AND WAVEGUIDES

7-1 INTRODUCTION

Another method of guiding electrical energy besides that of two-wire transmission is the single conductor waveguide which is employed at the higher frequencies, above 3 GHz or so, to obtain larger bandwidths and lower signal attenuation. The waveguides usually take the shape of a rectangular or cylindrical structure, although at times other irregular shapes are used.

Associated with the voltage and current along a transmission line is an electric and magnetic field. The current probe employed in some clip-on ammeters, for instance, monitors the surrounding magnetic field around a conductor, which, in turn, bears a direct relationship to the current passing through the conductor. The fields associated with the more common types of transmission lines are illustrated in Section 1-2. It is possible to treat the electrical properties of a transmission line on either a voltage–current or electromagnetic field basis, with the later being much more involved mathematically. When propagation in single conductor waveguides or in free space is studied, the field approach is normally employed. Even then, the currents in the walls of a waveguide may also be of interest, in particular when slots are machined in the walls to permit radiation.

For time-varying fields, both the electric and magnetic fields are always present. This is not necessarily the case for the static situation. The relationships between the electric and magnetic fields and between these fields and the environment (boundary conditions) are given by Maxwell's equations. In these three-dimensional situations, two space variables must be added to the concepts employed in transmission-line theory, causing the solutions to take on a more complex format. The solutions to Maxwell's equations are vector quantities, and thus a knowledge of vector analysis is required. Since vector analysis is not assumed as a prerequisite for readers of this book, we will not employ Maxwell's equation as a starting point. Many textbooks written on electromagnetic wave theory can be referred to if the reader wishes to follow this more rigorous approach. In this and following chapters, we will attempt to obtain a grasp of field theory by using an analogy with what we have learned with respect to the voltage and current along a transmission line. We will first explore the nature of a plane electromagnetic (EM) wave and then study the fields in a rectangular and circular waveguide structure.

7-2 UNIFORM PLANE WAVE

The two components of any EM wave are the electric field (**E**) and the magnetic field (**H**). These are both vector quantities, as indicated by the boldface type, having magnitude, phase angle, and direction. In free space the EM wave travels at the speed of light (3×10^8 m/s). In other media it travels at a velocity of

$$v = \frac{c}{\sqrt{\epsilon_r \mu_r}}$$

where c = velocity of light

ϵ_r = relative dielectric constant of the media

μ_r = relative permeability constant of the media

Because of its simplicity, we shall first discuss the uniform plane wave. The plane wave is a good approximation to the field observed at a point a fairly large distance from a transmitting antenna. Also, as we shall discuss in more detail in Section 7-5, the wave propagating down a rectangular waveguide can be assumed to be made up of two plane waves, zigzaging down the guide.

The unique feature of the plane wave is that its power density remains constant as it propagates. The plane wave has no variation of the field in the direction perpendicular to the direction of propagation. A graphical illustration of a plane wave is shown in Fig. 7-1. Such a wave has the electric and magnetic field vector always at right angles to each other and they are both also at right angles to the direction of propagation.

The direction of propagation is given by the right-hand-screw rule. Turn the

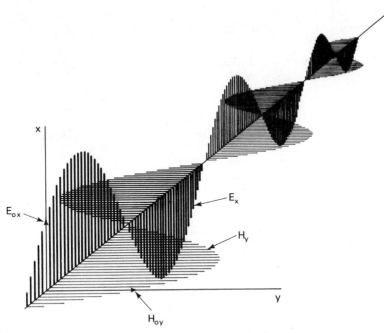

FIG. 7-1 Plane electromagnetic wave.

electric field intensity vector onto the magnetic field intensity vector through 90°, and the direction that a screw would move (whose axis lies in the direction of propagation) is the direction of propagation. Figure 7-2 illustrates this procedure. For the plane wave illustrated in Fig. 7-1, the fields can be expressed in a manner similar to that of the traveling waves on a transmission line. Thus,

$$E_x(z, t) = \text{Re}[E_{0x}e^{j(\omega t - \beta z)}] \tag{7-1}$$

$$H_y(z, t) = \text{Re}[H_{0y}e^{j(\omega t - \beta z)}] \tag{7-2}$$

$$E_y = E_z = H_x = H_z = 0$$

where E_x represents the x component of the electric field and H_y represents the y

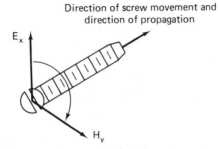

FIG. 7-2 Vector relationship of the electric **E**, magnetic **H**, and direction of propagation.

component of the magnetic field. Both of these equations represent fields traveling in the $+z$ direction. The **E** and **H** fields are vector quantities and have peak phases of $E_{0x}e^{-j\beta z}$ and $H_{0y}e^{-j\beta z}$, respectively. In the x and y directions there are no variations in the field whatsoever. The relationship between the electric and magnetic field is also unique and is given by

$$\frac{|E|}{|H|} = \eta \tag{7-3}$$

where η is called the intrinsic or characteristic impedence of the medium.

In free space

$$\eta = \eta_0 = \sqrt{\frac{\mu_0}{\epsilon_0}} = 120\pi = 377 \ \Omega \tag{7-4}$$

More generally,

$$\eta = \sqrt{\frac{j\omega\mu}{\sigma + j\omega\epsilon}} \tag{7-5}$$

where σ is the conductivity of the medium (S/m).

In the case where the conductivity is finite, there is also an attenuation constant. In this case, the fields are expressed by

$$E_x(z, t) = \text{Re}(E_{0x}e^{j\omega t - \gamma z}) \tag{7-6}$$

$$H_y(z, t) = \text{Re}(H_{0y}e^{j\omega t - \gamma z}) \tag{7-7}$$

where γ is the propagation constant and is given by

$$\gamma = \alpha + j\beta = \sqrt{j\omega\mu(\sigma + j\omega\epsilon)} \tag{7-8}$$

As can be seen by the equations thus far, the theory built up around transmission lines carries over to the plane-wave problem by direct analogy. The analogies between the transmission line and the plane wave are tabulated in Table 7-1. It is easier to use these analogies than to relearn a completely new set of formulas.

EXAMPLE 7-1

Let us obtain the propagation constant through mica, which has an extremely small conductivity (assume it to be zero). For the transmission-line analogy,

$$\gamma = \sqrt{(R + j\omega L)(G + j\omega C)}$$

Substituting in for L and C, μ and ϵ, respectively (in some texts τ is given as an analogy for R, but we will not employ it),

$$\gamma = \sqrt{(j\omega\mu)(j\omega\epsilon)} = j\omega\sqrt{\mu\epsilon} \tag{7-9}$$

Hence, in mica

$$\alpha = 0 \quad \text{and} \quad \beta = \omega\sqrt{\mu\epsilon} \tag{7-10}$$

The phase velocity in mica is given by

$$v_p = \frac{\omega}{\beta} = \frac{1}{\sqrt{\mu\epsilon}} \tag{7-11}$$

TABLE 7-1 Transmission Line/Plane Wave Analogy

Transmission-line quantity	Symbol and equation	Unit	Plane-wave quantity	Symbol and equation	Unit
Current	I	A	Magnetic field intensity	H	A/m
Voltage	V	V	Electric field intensity	E	V/m
Inductance per unit length	L	H/m	Permeability (inductivity)	μ	H/m
Capacitance per unit length	C	F/m	Dielectric constant (capacitivity)	ϵ	F/m
Resistance per unit length	R	Ω/m			
Conductance per unit length	G	S/m	Conductivity	σ	S/m
Characteristic impedance	$Z_0 = \sqrt{\dfrac{R+j\omega L}{G+j\omega C}}$	Ω	Intrinsic impedance	$\eta = \sqrt{\dfrac{j\omega\mu}{\sigma+j\omega\epsilon}}$	Ω
Propagation constant	$\gamma = \alpha + j\beta$ $= \sqrt{(R+j\omega L)(G+j\omega C)}$	Np/m and rad/m	Propagation constant	$\gamma = \alpha + j\beta$ $= \sqrt{j\omega\mu(\sigma+j\omega\epsilon)}$	Np/m and rad/m
Phase velocity	$v_p = \dfrac{\omega}{\beta}$	m/s	Phase velocity	$v_p = \dfrac{\omega}{\beta}$	m/s
Incident power	$P^+ = \dfrac{|V^+|^2}{Z_0}$	W	Incident power density	$\mathcal{P}^+ = \dfrac{|E^+|^2}{\eta}$	W/m²

In mica, as for most other dielectrics $\mu_r = 1$. Therefore,

$$v_p = \frac{c}{\sqrt{\epsilon_r}}$$ (7-12)

EXAMPLE 7-2

Consider a wave in air impinging normally on a finite slab of polyethylene 1 cm thick as shown in Fig. 7-3(a). Assume the polyethylene to be lossless and have an ϵ_r of 2.25. Find

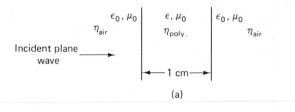

(a)

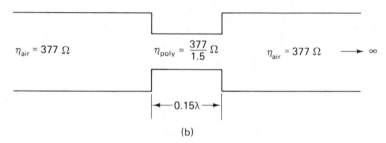

(b)

FIG. 7-3 Transmission-line analogy for Example 7-2.

the reflection coefficient at the incident face of the sheet, the reflected and incident power densities, if the incident electric field intensity is 10 V/m (rms) and the frequency of operation is 3 GHz.

Solution:

$$\eta_{\text{air}} = \sqrt{\frac{\mu_0}{\epsilon_0}} = 377 \,\Omega$$

$$\eta_{\text{poly}} = \sqrt{\frac{j\omega\mu}{j\omega\epsilon}} = \sqrt{\frac{\mu_0}{\epsilon_r\epsilon_0}} = \frac{377}{1.5} \,\Omega$$

In order to determine the width of the dielectric slab in wavelengths so that the Smith chart can be employed, the phase constant will first be determined.

$$\gamma_{\text{poly}} = \sqrt{j\omega\mu\, j\omega\epsilon} = j\omega\sqrt{\mu_0\epsilon_r\epsilon_0} = j30\pi$$

$$\alpha_{\text{poly}} = 0 \quad \text{and} \quad \beta_{\text{poly}} = 30\pi \text{ rad/s}$$

Since

$$\lambda = \frac{2\pi}{\beta}$$

$$\lambda_{\text{poly}} = \frac{2\pi}{30\pi} = \frac{1}{15} \text{ m}$$

Thus, the thickness of the slab in wavelengths is

$$l = \frac{0.01}{1/15} = 0.15 \text{ wavelength}$$

Since the region to the right of the slab is assumed to be infinitely long, the load impedance as seen at the right boundary of the slab will be 377 Ω. The equivalent transmission line thus can be reduced to that shown in Fig. 7-4.

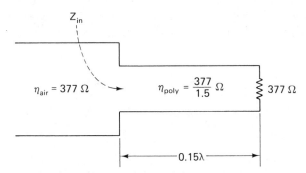

FIG. 7-4 Reduced equivalent circuit for Example 7-2.

The input impedance (Z_{in}) can now be obtained by plotting the normalized impedance $377/(377/1.5) = 1.5$ on the Smith chart and rotating 0.15λ toward the generator. This results in an input impedance of

$$Z_{\text{in}} = \frac{377}{1.5}(0.82 - j.32)$$

$$= 206 - j\,80\Omega$$

The reflection coefficient at the incident face of the sheet is therefore

$$\Gamma = \frac{Z_{\text{in}} - \eta_{\text{air}}}{Z_{\text{in}} + \eta_{\text{air}}}$$

$$= 0.32\underline{/-147°}$$

The reflected wave has an amplitude of

$$|E^-| = |\Gamma||E^+| = (0.32)(10) = 3.2 \text{ V/m}$$

Thus,

$$\mathcal{P}^+ = \frac{(10)^2}{377} = 0.266 \text{ W/m}^2$$

$$\mathcal{P}^- = \frac{(3.2)^2}{377} = 0.027 \text{ W/m}^2$$

The power transmitted through the slab must be $\mathcal{P}$ transmitted $= \mathcal{P}^+ - \mathcal{P}^- = 0.239$ W/m².

7-3 CONDUCTORS AND DIELECTRICS

Let us consider again for a moment the lossy material shown in Fig. 1-5, which has a shunt conductivity of σ and a shunt capacitivity (dielectric constant) of ϵ. If σ is very small compared to $\omega\epsilon$, we consider the material to be a good *dielectric*. If, however, σ is very large compared to $\omega\epsilon$, the material is considered to be a good conductor. The properties of dielectrics are usually given in terms of the dielectric constant and the ratio $\sigma/\omega\epsilon$. The ratio $\sigma/\omega\epsilon$ is also known as the *dissipation factor*, *D*, of the dielectric.

Let us now consider a plane wave just entering such a lossy sheet of material as shown in Fig. 7-5. If the lossy material is a good conductor ($\sigma/\omega\epsilon \gg 1$), the

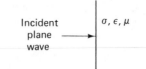

Incident plane wave → | σ, ϵ, μ

FIG. 7-5 Plane wave incident upon a lossy material.

expression for the propagation constant may be expressed by

$$\gamma = \sqrt{(j\omega\mu)(\sigma + j\omega\epsilon)} = \sqrt{j\omega\mu\sigma\left(1 + \frac{j\omega\epsilon}{\sigma}\right)}$$

$$\approx \sqrt{j\omega\mu\sigma} = \sqrt{\omega\mu\sigma}\,\underline{/45^\circ} \tag{7-13}$$

$$= \sqrt{\frac{\omega\mu\sigma}{2}} + j\sqrt{\frac{\omega\mu\sigma}{2}}$$

from which can be concluded that

$$\alpha = \beta = \sqrt{\frac{\omega\mu\sigma}{2}} \tag{7-14}$$

In a good conductor, therefore, both α and β are large. Since the attenuation increases with frequency, at microwave and radio frequencies the rate of attenuation in a good conductor is very large. Thus, for good shielding, both high μ and high σ are required. From equation (7-14) we can gather that to obtain good shielding, a good conductor (high σ) with a high permeability constant should be employed.

A term frequently employed when dealing with waves experiencing high attenuations is *depth of penetration*. This term is analogous to the time-constant term in *RL* and *RC* circuits. If at the surface of a plane conductor a field exists having a magnitude of E_{0x}, the field inside will have a magnitude of $|E_x| = |E_{0x}|e^{-\alpha z}$. This is shown in Fig. 7-6. The depth of penetration (also frequently called *skin effect*) is defined as that depth at which the wave is attenuated to or 37% of its value at the surface. This occurs when the magnitude of the field reaches $|E_{0x}|e^{-1}$.

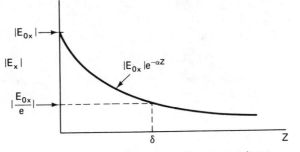

FIG. 7-6 Field strength near the surface in a conductor.

Using the symbol δ for depth of penetration, it will have a value of

$$\alpha\delta = 1 \quad \text{or} \quad \delta = \frac{1}{\alpha} \tag{7-15}$$

which from equation (7-14) is equal to

$$\delta = \sqrt{\frac{2}{\omega\mu\sigma}} \tag{7-16}$$

Similar to the time-constant situation, it is generally assumed that after five δ's, the wave is considered to be attenuated to zero. The depth of penetration of Cu and Ag, the best known conductors, will now be obtained at 100 MHz and 10 GHz. The conductivities for Cu and Ag are usually taken as

$$\sigma_{Cu} = 5.8 \times 10^7 \text{ S/m}$$

$$\sigma_{Ag} = 6.2 \times 10^7 \text{ S/m}$$

The permeability and dielectric constants are the same as that in free space, that is,

$$\mu = \mu_0 \quad \epsilon = \epsilon_0$$

For Cu at 100 MHz,

$$\delta_{Cu} = \sqrt{\frac{2}{2\pi \times 10^8 \times 4\pi \times 10^{-7} \times 5.8 \times 10^7}} = 0.00661 \text{ mm}$$

For Cu at 10 GHz,

$$\delta_{Cu} = 0.000661 \text{ mm}$$

Similarly, for Ag:

$$\delta_{Ag} \text{ at 100 MHz} = 0.0064 \text{ mm}$$

$$\delta_{Ag} \text{ at 10 GHz} = 0.00064 \text{ mm}$$

Commercial waveguides have wall thicknesses which are generally much larger than these values in order to realize sufficient mechanical rigidity. The X-band waveguide (RG52/X), for instance, has a wall thickness of 0.05 in. $= 1.97$ mm, which is in the order of hundreds of skin depths thick.

It is because of the shielding effect of a conductor that radio waves cannot penetrate the body of an automobile. Often, copper screening is used to form a shielded room when tuning a sensitive receiver, especially in locations where out-side radio interference is present. In good conductors, where the current density (**J**) is related to the electric field intensity by the equation (in actuality, Ohm's law)

$$\mathbf{J} = \sigma\mathbf{E} \qquad (7\text{-}17)$$

the current follows the same pattern as that for the electric field. As the frequency is increased, most of the current flows near the surface of the conductor, which at extremely high frequencies or for almost perfect conductors can be considered as a current sheet.

7-4 BOUNDARY CONDITIONS

When determining a field configuration, the *boundary conditions* (the knowledge of the fields on a surface enclosing the region of interest) must be taken into account. To aid in the understanding of the fields in a waveguide, the general boundary conditions will be briefly introduced. The following conditions hold for the electric and magnetic fields at the surface of any boundary.

1. The *tangential component of the electric field* is continuous across the boundary. This can be mathematically stated as

$$E_{y1} = E_{y2} \qquad (7\text{-}18)$$

where E_{y1} and E_{y2} represent the components parallel to the surface, as indicated on Fig. 7-7.

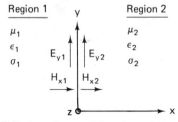

FIG. 7-7 Boundary conditions for an EM wave.

2. The *normal component of the magnetic flux density* is continuous across the boundary. Using the symbols indicated in Fig. 7-7, this can be stated as

$$\mu_1 H_{x1} = \mu_2 H_{x2} \qquad (7\text{-}19)$$

3. The *tangential component of the magnetic field* is continuous across a surface except at the surface of a perfect conductor. At the surface of a perfect conductor, it is discontinuous by an amount equal to the surface current per unit width. This could be written as (with region 2 being a perfect conductor)

$$H_{y1} = -J_z \qquad (7\text{-}20)$$

where J_z represents a current sheet flowing at the surface of the conductor in the z direction. The tangential component of **H**, the direction of the current density and the surface normal, are mutually perpendicular. As indicated in the previous section, current flows near the surface in a perfect conductor. This is even a good approximation for actual conductors at high frequencies. As an example, at point A in the coaxial line of Fig. 1-4, the tangential **H** field is in the vertical direction, the line normal to the surface is radial and inward, and the current is in the axial direction. This current flows near the inward surface of the outer conductor. At point B, where the surface normal is radial and outward, the current must flow in the opposite direction. Since the current flow is axial, a slot may be cut longitudinally in a coaxial line without altering the fields within the line, as in the case of the slotted line.

The significance of statement 1 for us lies in the case of a perfect conductor (silver and copper approach such a condition). Since in a perfect conductor a voltage gradient cannot exist, the electric field within it must also be zero. From statement 1 it can be concluded that there *cannot* be a tangential component of the electric field at the boundary. This condition is similar to the condition that a voltage cannot exist across a short circuit. If a good conductor is placed in an electromagnetic field, a standing wave must exist or else the entire field must be zero. A little later we will see this to be the case in any single conductor waveguide.

Statement 2 indicates that the magnetic lines of force are continuous or form closed loops. Thus, for a perfect conductor wherein both **E** and **H** are zero, there can be no normal component of the **H** field on either side of the conductor boundary.

Much more could be said about boundary conditions, which in the more advanced treatises on waveguide theory are expressed in a vector formula. Suffice it to say that these few brief statements will be of some aid in our understanding of the field configurations that will be encountered in waveguides.

7-5 RECTANGULAR WAVEGUIDE

Probably the most common type of waveguide presently employed is the *rectangular waveguide*, so named because of its shape. The walls are generally brass or aluminum and are structurally self-supporting.

Notation

When dealing with situations having rectangular symmetry, the rectangular coordinate system is employed for convenience and simplicity. In this system the three coordinates, x, y, and z, are perpendicular to each other as shown in Fig. 7-8. The rectangular waveguide is appropriately situated in the rectangular coordinate system.

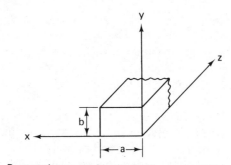

FIG. 7-8   Rectangular waveguide situated in a rectangular coordinate system.

Since electric and magnetic fields are vector quantities, the functional notation must indicate the component being described. We will denote all components by a subscript. Thus $E_x(x, y, z)$ denotes the x component of the electric field, which may vary with x, y, and z.

Later, when considering the cylindrical waveguide, the cylindrical coordinate system will be introduced.

Modes

Coaxial transmission lines commonly operate in what is called the *transverse electic magnetic* (TEM) *mode*. In this case both the electric and magnetic fields are perpendicular or transverse to the direction of propagation. No field component exists in the axial direction. In the case of single conductor waveguides (hollow pipes), this mode cannot exist; instead, either the TE (transverse electric) or the TM (transverse magnetic) mode can be energized. Within either grouping, a number of configuration or modes can exist, either separately or simultaneously. Also, both modes can exist together, although this is generally avoided if possible.

In many European publications the designations **H** and **E** are used rather than TE and TM. These designations refer to the axial components that are present rather than to the transverse components. In the case of the TE configuration, the electric field is transverse to the direction of propagation. Using the waveguide orientation as given in Fig. 7-8, this means that there may be an x and y component of the electric field **E**, but there may not be a component of the electric field in the direction of propagation (z direction). There is, however, a magnetic field component in the z direction, and this is the feature that the Europeans use in their designations. On the other hand, in the TM configuration, the magnetic field must be transverse or perpendicular to the direction of propagation. In this case there is an axial component of the electric field present.

To denote the mode being propagated, two subscripts are added to these designations, the pair of indexes m and n; thus, the designations $\mathrm{TE}_{m,n}$ ($\mathbf{H}_{m,n}$) and $\mathrm{TM}_{m,n}$ ($\mathbf{E}_{m,n}$) are used. For example, the TE_{23} mode represents a transverse electric field having an $m = 2$ and $n = 3$. Later we will see the significance of the mode numbers.

The general expressions for the $\mathrm{TE}_{m,n}$ and $\mathrm{TM}_{m,n}$ modes in a rectangular waveguide are given in Appendix C. These equations are derived from Maxwell's equations in many basic electromagnetic field texts. In these notes we will consider in some detail the TE_{10} mode which is often called the dominant mode (defined as the lowest frequency mode that can exist in the waveguide). The TE_{10} mode is the most common mode used in rectangular waveguides. To determine the corresponding field configurations, we set $m = 1$ and $n = 0$ in equations (C-1) and (C-3) of Appendix C. Thus, for the TE_{10} mode:

$$E_x = 0 \tag{7-21a}$$

$$E_y = \frac{-j60\pi c f H_0}{a\epsilon_r f_c^2} \sin \frac{\pi x}{a} e^{-j\beta_g z} \tag{7-21b}$$

$$E_z = 0 \tag{7-21c}$$

$$H_x = \frac{jc\sqrt{f^2 - f_c^2}}{2a\sqrt{\mu_r \epsilon_r} f_c^2} H_0 \sin \frac{\pi x}{a} e^{-j\beta_g z} \tag{7-21d}$$

$$H_y = 0 \tag{7-21e}$$

$$H_z = H_0 \cos \frac{\pi x}{a} e^{-j\beta_g z} \tag{7-21f}$$

where

$$f_c = \frac{c}{2a\sqrt{\mu_r \epsilon_r}} \tag{7-22}$$

The arbitrary constant H_0 in the equations depends upon the source supplying energy to the waveguide. a is the width of the waveguide and μ_r and ϵ_r are the relative permeability and dielectric constants of the medium inside the waveguide.

Equation (7-21) informs us that there are no x and z components of the electric field (only a y component exists) and that there is no y component of the magnetic field intensity. The y component of E also varies in magnitude sinusoidally along x, with this component being zero at the waveguide boundaries $x = 0$ and $x = a$. All these equations represent traveling waves traveling in the plus Z direction having a phase constant equal to β_g. This phase constant is related to the waveguide wavelength by

$$\beta_g = \frac{2\pi}{\lambda_g} \tag{7-23}$$

To obtain some physical understanding of waveguide wavelength, λ_g, as opposed to free-space wavelength, λ, we will consider another way of looking how the fields are set up in the rectangular waveguide structure. Before doing this, let us first draw the field configurations of the TE_{10} mode. This is done in Fig. 7-9, where the lines follow the directions of the field vectors. The space density of the lines indicate the relative strength of the field. The fields set up in a rectangular waveguide propagating the TE_{10} mode can also be considered to be made up of two plane waves traveling in slightly different directions. The angle of travel depends upon the wavelength and upon the waveguide width dimension.

Consider the two sinusoidal plane waves of the same amplitude and frequency shown in Fig. 7-10(a). The areas around the crest maxima are shown by solid

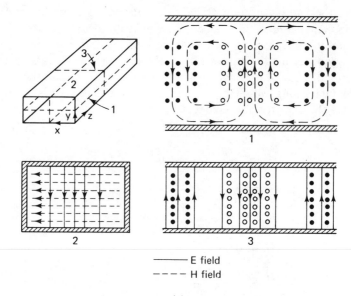

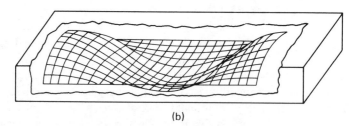

FIG. 7-9 Field configuration for the TE_{10} mode in a rectangular waveguide. (a) Closeness of line spacing indicates the strength of the field. Dots represent field lines coming out of the plane of the paper. Small circles represent lines going into paper. [From S. Ramo, J. R. Whinnery, and T. Van Duzer, *Fields and Waves in Communication Electronics* (New York: John Wiley & Sons, Inc., 1965), Table 8.02, p. 423.] (b) Three-dimensional view of the electric field.

lines, whereas the dotted lines represent the locations of the crest minima. At the points where the electric intensity of one wave is equal but opposite to the electric intensity of the other wave, the field cancels. In Fig. 7-10(a) this cancellation of the electric field occurs at the location of the horizontal lines. Since no field is present there, a conductor or waveguide wall could be placed at this location without any averse affects. By completing the enclosure with top and bottom plates, the rectangular waveguide is formed. In regions away from the sidewalls, the sum of the fields is not zero and the resultant coincides with the field configuration shown in Fig. 7-9. In actual practice there is but one wave, each wave being the reflection of the other.

Figure 7-10(b) illustrates the zigzaging pattern of the rays. This way of look-

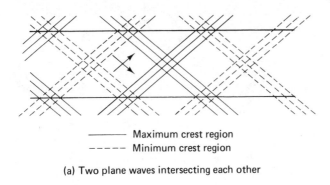

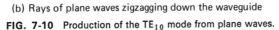

——— Maximum crest region
----- Minimum crest region

(a) Two plane waves intersecting each other

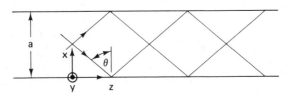

(b) Rays of plane waves zigzagging down the waveguide

FIG. 7-10 Production of the TE_{10} mode from plane waves.

ing at wave propagation aids in the understanding of such concepts as phase and group velocity, cutoff frequency, and waveguide wavelength.

7-6 PHASE AND GROUP VELOCITY

In the case of a rectangular or cylindrical waveguide, the wave or phase velocity (the velocity of a constant phase front) is always greater than the free-space velocity. The group velocity (the velocity of energy propagation in the $\pm z$ direction), on the other hand, is always less than the free-space velocity.

To obtain a clearer understanding of phase velocity, consider a water wave approaching a shore line. Figure 7-11 shows a wave with a velocity $c/\sqrt{\mu_r \epsilon_r}$ approaching a shoreline at an angle θ to the normal. The velocity of the waves can be determined by measuring the distance between successive crests (λ) and noting the frequency (f) with which the crest passes a given observation point. The velocity of the incoming wave can be expressed as

$$\frac{c}{\sqrt{\mu_r \epsilon_r}} = \lambda f$$

If an observer on the shore, however, observes the distance between crests along the shoreline (λ_z), the velocity that he appears to observe will be

$$v_p = \lambda_z f$$
$$= \frac{\lambda}{\sin \theta} f = \frac{c/\sqrt{\mu_r \epsilon_r}}{\sin \theta} \tag{7-24}$$

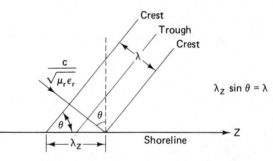

FIG. 7-11   Water wave approaching a shoreline.

This velocity, called the phase velocity, is always greater than $c/\sqrt{\mu_r \epsilon_r}$. Also, the wavelength along the shoreline is greater than the wavelength along the direction of signal wave travel.

The group velocity, or velocity of energy propagation along the guide, is equal to

$$v_g = \frac{c}{\sqrt{\mu_r \epsilon_r}} \sin \theta \qquad (7\text{-}25)$$

This quantity is always less than $c/\sqrt{\mu_r \epsilon_r}$. We can note that

$$v_p \cdot v_g = \left(\frac{c}{\sqrt{\mu_r \epsilon_r}} \right)^2 \qquad (7\text{-}26)$$

In the case of the hollow waveguide the velocity c is the velocity of light. The zigzag path of the TE_{10} (as well as the other more complex modes) causes the observed wavelength to be different from that of the free-space wavelength. The phase velocity in waveguides depends upon the frequency of operation and in general is given by

$$v_p = \frac{c}{\sqrt{\mu_r \epsilon_r} \sqrt{1 - (f_c/f)^2}} \qquad (7\text{-}27)$$

where μ_r and ϵ_r = relative permeability and permittivity constants of the material contained in the waveguide

f_c = cutoff frequency

f = frequency of operation

The corresponding waveguide wavelength is given by

$$\lambda_g = \frac{v_p}{f} = \frac{c}{\sqrt{\mu_r \epsilon_r} \sqrt{f^2 - f_c^2}} \qquad (7\text{-}28)$$

where λ_g is equivalent to the shoreline wavelength of Fig. 7-11.

The cutoff frequency, f_c, the frequency below which no wave propagation occurs, arises from the phenomenon that as the frequency of operation is decreased, the zigzag angle (θ of Fig. 7-10) decreases. At the frequency f_c, the waves merely bounce back and forth between the walls, and no wave motion occurs parallel to the z axis. The general equation for f_c is given in Appendix C, which for the TE_{10} mode reduces to equation (7-22).

For an air-filled (or evacuated) waveguide, the cutoff frequency for the TE_{10} mode is

$$f_{c10} = \frac{c}{2a} \qquad (7\text{-}29)$$

Restating equation (7-29) in free-space wavelength (λ) terms,

$$2a = \frac{c}{f_{c10}} = \lambda_c \qquad (7\text{-}30)$$

where λ_c is called the free-space cutoff wavelength (c/f_{c10}). Figure 7-12 indicates the variation of phase and group velocity with the frequency of operation.

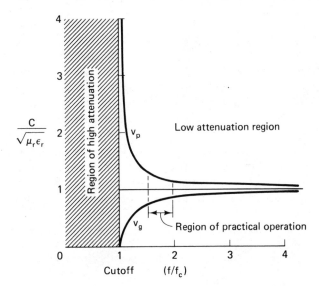

FIG. 7-12 Variation of phase velocity and group velocity with frequency.

From the discussion so far, it should be clear that a voltage or current pulse cannot be sent down a waveguide to determine the location of irregularities, and so on, as is done in time-domain reflectometry. The velocity of propagation varies with the many frequency components and "dispersion" takes place where the returned wave is greatly distorted. In the case of radar, where a pulse is sent out, a pulse of RF must be transmitted. The velocity that this pulse experiences is the group velocity and it is given by [see equation (7-26)]

$$v_g = \frac{(c/\sqrt{\mu_r \epsilon_r})^2}{v_p} \qquad (7\text{-}31)$$

In Fig. 7-12 the region of practical operation is marked out. For a signal of some bandwidth operating below this region, a large dispersion will result and the associated distortion can become unreasonably high. Waveguide runs should be kept at a minimum to reduce dispersion. At frequencies above this marked-out region, however, higher modes can be set up, which can result in inefficient transmission. Appendix D gives a table of several types of rigid rectangular waveguides, which includes a recommended operating range.

7-7 HIGHER-ORDER MODES IN RECTANGULAR WAVEGUIDES

It is generally desired to propagate only one mode in a waveguide, as it is difficult to design coupling devices that will be matched for several modes; each mode may have a different group velocity, and so on. Mode jumping can also occur, as the wave has a tendency to decouple itself from a load if possible.

The commonest method employed to avoid these problems is to operate the waveguide in its dominant or lowest-frequency mode and design the size of the waveguide such that the signal frequency is below the cutoff frequencies of all except this lowest mode. The wave then cannot propagate in any other mode.

The cutoff frequency for the dominant mode is given by equation (7-29) and is seen to be only dependent upon the waveguide width a. It is assumed in this case that b is less than a. For the higher-order modes, the cutoff frequency can be determined from Appendix C, equation (C-3). If we set the ratio of b/a to $\frac{1}{2}$, the TE_{01} and TE_{20} modes both have cutoff frequencies equal to c/a, which is twice the TE_{10} cutoff frequency. As the b/a ratio is increased, to 1, for example, the cutoff frequencies of the TE_{10} and TE_{01} modes become equal. In this case these two modes could propagate simultaneously.

In principle the mode set up in the waveguide is dependent upon the nature of coupling employed. In practice, irregularities in the waveguide and at points of discontinuity such as bends, coupling tuners, and so on, cause other modes to be set up. Figure 7-13 shows the relative cutoff frequencies for several rectangular waveguide modes.

To obtain some feeling for higher-order modes, one can relate to the m, n

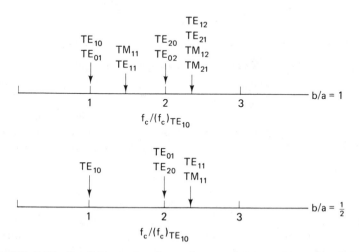

FIG. 7-13 Cutoff frequencies of several rectangular waveguide modes relative to the TE_{10} mode cutoff frequency. [From S. Ramo, J. R. Whinnery, and T. Van Duzer, *Fields and Waves in Communication Electronics* (New York: John Wiley & Sons, Inc., 1965), Fig. 8.02b, p. 423.]

integers the variations in field amplitudes. The *m* integer relates to the number of half-sinusoid amplitude variations that occur in the field strength as one moves along the *a* dimension of the waveguide. The *n* integer relates similarly to the *b* dimension. With this and the boundary condition that the tangential electric field must be zero at the waveguide surface, it becomes quite easy to sketch fields for almost any mode. Figure 7-14 shows the field configuration for the TE_{20}, TE_{11}, and TE_{21} modes.

The field configurations for the TM mode are similar to those of the TE except that in this case the tangential field is totally transverse to the direction of propagation. Also, there is an axial component of the electric field present. The *m, n* integers have the same connotations as before. The TM_{11} and TM_{21} modes are shown in Fig. 7-15.

7-8 ATTENUATION IN RECTANGULAR WAVEGUIDES

The equations for the fields in a rectangular waveguide given thus far assume no attenuation when the operating frequency is kept above the cutoff frequency. For the practical case where some loss is present as a result of the small resistance of the

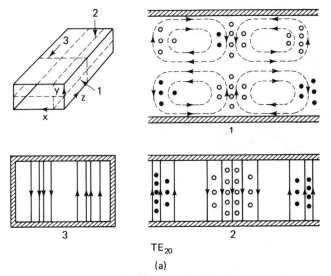

TE_{20}

(a)

FIG. 7-14 Field configurations for several TE modes in a rectangular waveguide. Closeness of the line spacing indicates the strength of the field. Dots represent field lines coming out of the plane of the paper. Small circles represent lines going into paper. [From S. Ramo, J. R. Whinnery, and T. Van Duzer, *Fields and Waves in Communication Electronics* (New York: John Wiley & Sons, Inc., 1965, Fig. 8.02b, p. 423.]

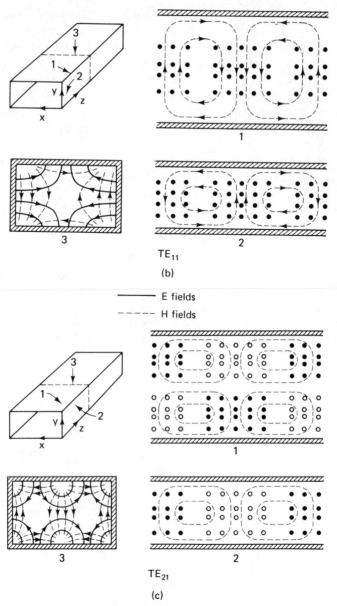

E fields
- - - - H fields

TE_{11}

(b)

TE_{21}

(c)

FIG. 7-14 (Continued)

waveguide walls, the fields are not appreciably distorted from the case presented. If a dielectric other than air is used, which is seldom the case, some dielectric losses will also be present.

Figure 7-16 shows the attenuation curve as a function of frequency for a typical 1 in. × 2 in. brass rectangular waveguide. It can be noted that the attenuation increases with frequency for all modes at the higher frequencies, with a minimum occurring at some frequency above the cutoff frequency.

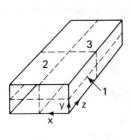

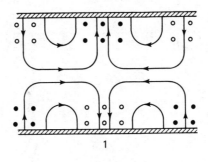

1

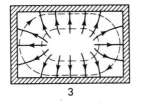

3

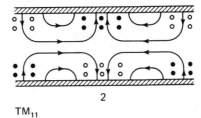

2

TM_{11}

(a)

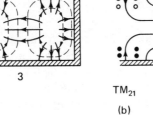

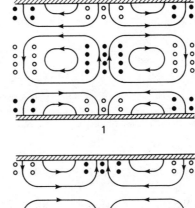

1

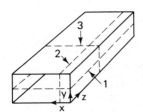

3

2

TM_{21}

(b)

———— E fields

– – – – – H fields

FIG. 7-15 Field configurations for two TM modes in a rectangular waveguide. Closeness of line spacing indicates the strength of the field. Dots represent field lines coming out of the plane of the paper. Small circles represent lines going into paper. [From S. Ramo, J. R. Whinnery, and T. Van Duzer, *Fields and Waves in Communication Electronics* (New York: John Wiley & Sons, Inc., 1965), Fig. 8.02b, p. 423.]

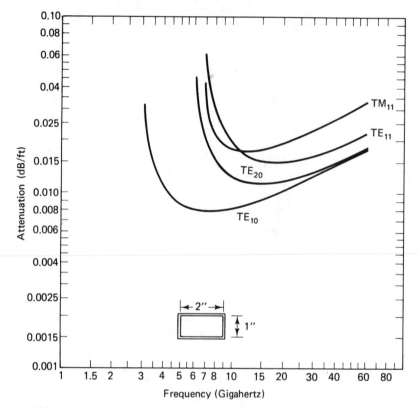

FIG. 7-16 Attenuation due to copper losses in a 2 in. × 1 in. rectangular brass waveguide. [From E. C. Jordan and K. G. Balmain, *Electromagnetic Waves and Radiating Systems* (Englewood Cliffs, N.J.: Prentice-Hall, Inc., 1968), p. 271.]

7-9 CHARACTERISTIC IMPEDANCE IN RECTANGULAR WAVEGUIDES

Similar to the characteristic impedance defined for a transmission line, a characteristic wave impedance may be defined for a waveguide. The characteristic wave impedance is defined in terms of the field components that lie in a plane transverse to the direction of propagation. The reason for employing only these components is that the component parallel to the waveguide axis results in a component of power flow across the guide as the direction of power flow is perpendicular to the field vectors. Since our interest centers on the impedance as seen when looking in the direction of propagation, that is, along the z axis, only the x and y components are involved.

The characteristic wave impedance is thus defined as

$$Z_0 = \frac{E_{\text{transverse}}}{H_{\text{transverse}}} = \frac{\sqrt{E_x^2 + E_y^2}}{\sqrt{H_x^2 + H_y^2}} \tag{7-32}$$

By substituting the expression as for E_x, E_y, H_x, H_y, given by equation (C-1) in

Appendix C, we obtain for the $\text{TE}_{m,n}$ modes the characteristic impedance

$$Z_{0(\text{TE}_{m,n})} = \frac{\eta}{\sqrt{1 - (f_c/f)^2}} \tag{7-33}$$

Similarly, substituting the expression as given by equations (C-2) in Appendix C for the $\text{TM}_{m,n}$ mode cases, we obtain

$$Z_{0(\text{TM}_{m,n})} = \eta\sqrt{1 - (f_c/f)^2} \tag{7-34}$$

A sketch of $Z_{0(\text{TE}_{m,n})}$ and $Z_{0(\text{TM}_{m,n})}$ versus frequency is given in Fig. 7-17.

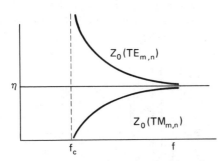

FIG. 7-17 Characteristic wave impedance as a function of operating frequency for the *TE* and TM rectangular waveguide modes.

Although the characteristic wave impedance is an interesting concept, it is very seldom used in practice, as it is very difficult to measure. When considering impedance-matching problems in waveguides, the Smith chart is usually employed after the null locations with respect to the load and the VSWR are measured. Impedances or admittances are then stated in terms of normalized values.

One note of caution should be raised at this point. An "open-ended" wave-guide, owing to its tendency to radiate its signal, is not a true open circuit. A short can still be obtained by clamping a flat metal plate over the end of the waveguide. An open can be simulated by attaching a quarter-wave section of waveguide to such a short (the length must be one-fourth of the waveguide wavelength).

A matched load is frequently obtained by mounting a length of power-absorbing material in the space near the closed end of the waveguide. Figure 7-18

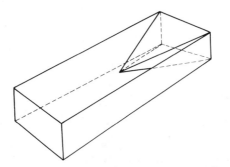

FIG. 7-18 Low-power waveguide matched pyramidal load.

shows such a structure. In large power applications, heat-radiating fins and water-cooled loads become necessary.

7-10 COUPLING TO WAVEGUIDES

The two most common ways of coupling fields from a coaxial line to a waveguide are the simple probe coupler and loop coupler. The *probe coupler* couples to the electric field and is usually about a quarter-wave long. This coupler is frequently used to couple a klystron source to a waveguide or as a transition from a coaxial line to a waveguide. Figure 7-19 shows such a coupler. The shorting plate approxi-

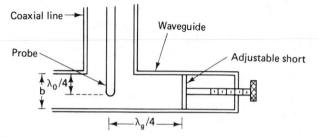

FIG. 7-19 Probe coupling from a coaxial line to a rectangular waveguide (TE_{10} mode).

mately $\lambda_g/4$ from the probe is often adjustable for tuning at various frequencies. The probe is centered in the waveguide to couple at the highest electric-field-intensity location. Figure 7-20 shows probe coupling to some of the other modes. The probe in all cases is parallel to the electric field vector.

Loop coupling couples to the magnetic field rather than to the electric field. This coupling is used to couple to one of the cavities in a magnetron tube for instance. Figure 7-21 indicates loop coupling to the TE_{10} mode. It is important that

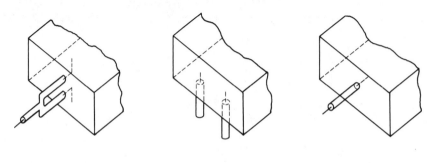

TE_{11} TE_{20} TM_{11}

FIG. 7-20 Probe coupling to the TE_{11}, TE_{20}, and TM_{11} modes.

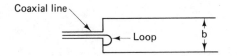

Coaxial line

← Loop   b

FIG. 7-21   Loop coupling to the TE_{10} mode in a rectangular waveguide.

these coupling devices are properly matched to the waveguide, and vice versa. This is often done by approximate design of the probe or loop, or by employing tuners.

7-11 CIRCULAR WAVEGUIDES

Circular waveguides are also frequently employed because they are easier to manufacture than rectangular ones and are also easier to join together. Some of the modes have the advantage of being axially symmetrical, allowing rotation of the waveguide (a rotating joint in a scanning antenna) without upsetting the field patterns. This waveguide is larger in area for the same cutoff frequency than the rectangular waveguide and does have the problem of polarization changes due to irregularities and bending of the waveguide.

Another very significant advantage of one family of modes in circular waveguides, the TE_{0l} modes, is that the attenuation coefficient decreases as the frequency of operation is increased. This is just the opposite characteristic of coaxial and parallel wire lines or rectangular waveguides, where the *skin effect* causes the conductor losses to increase in frequency.

To obtain more bandwidth and more spectrum space, the common carriers are looking toward increasingly higher operating frequencies. Since the wave propagation through the atmosphere is severely attenuated and scattered by rain, water vapor, and oxygen absorption at frequencies above 10 GHz, it appears that an enclosed waveguide is required. The circular electric wave modes (TE_{0l} modes) seem to provide such an alternative.

When dealing with cylindrically symetric systems, the cylindrical coordinate system is employed for convenience. Figure 7-22 shows a circular waveguide superimposed in a cylindrical coordinate system. In this coordinate system the field components are specified in terms of the p (radial), ϕ (angular), and z (axial) directions.

As in rectangular waveguides, the TE_{nl} (transverse electric) and TM_{nl} (transverse magnetic) modes can propagate. The subscripts n and l are also used here where the integer n now represents the number of full-wave intensity variations around the circumference around the waveguide, and l represents the number of half-wave intensity variations radially out from the center of the waveguide.

Whenever cylindrical symmetry is involved, one can expect to express the fields in terms of Bessel functions. The fields for the TE_{01} mode are given by equations (7-37), which contain the Bessel function of the first kind of order zero (J_0). As a result of this, the cutoff frequencies do not have the simple form as for rectan-

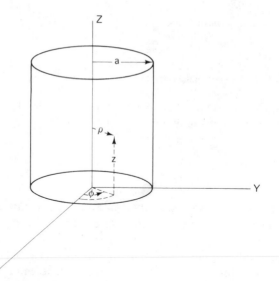

FIG. 7-22 Circular waveguide located in a cylindrical coordinate system.

gular waveguides, but involve the "zeros" of the Bessel functions. For example, the dominant TE_{11} mode (lowest frequency mode) has a cutoff frequency of

$$f_c(TE_{11}) = \frac{1.841}{2\pi a\sqrt{\mu\epsilon}} \tag{7-35}$$

where a is the radius of the circular waveguide in meters.

The cutoff frequency for the TE_{01} circular electric mode is given by

$$f_c(TE_{01}) = \frac{3.832}{2\pi a\sqrt{\mu\epsilon}} \tag{7-36}$$

Figure 7-23 shows the cutoff frequencies of the lower-order modes in circular wave-

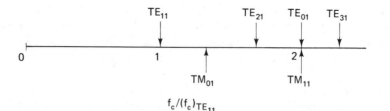

FIG. 7-23 Cutoff frequencies of modes relative to f_c (TE_{11}).

guides. The field patterns for some of the more frequently used modes are shown in Fig. 7-24.

The TE_{11} mode has been the most common of these modes, as it can be readily coupled to the TE_{10} mode in the rectangular guide. The rotary-vane type of waveguide attenuator (Table 7-3) for instance makes use of this mode. The TE_{11} mode is also able to support a circularly polarized TE wave.

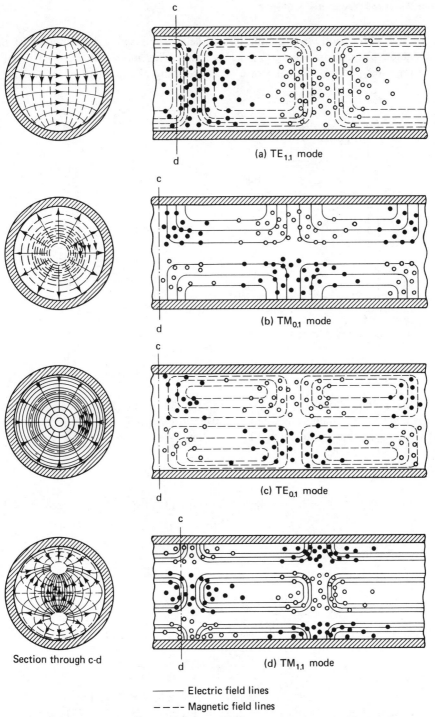

Section through c-d

(a) TE$_{1,1}$ mode

(b) TM$_{0,1}$ mode

(c) TE$_{0,1}$ mode

(d) TM$_{1,1}$ mode

——— Electric field lines
– – –– Magnetic field lines

FIG. 7-24 Field patterns of common modes in circular waveguides. [From A. B. Bronwell and R. E. Beam, *Theory and Application of Microwaves* (New York: McGraw-Hill Book Company, 1947), p. 336.]

The TM_{01} mode is a circularly symmetric mode, allowing rotation of the waveguide without disturbing the field pattern. This mode can be used when a rotating joint is used, as in the case of the rotating antenna shown in Fig. 7-25.

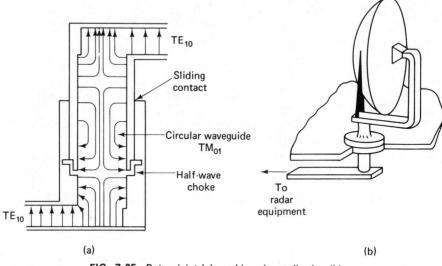

(a)   (b)

FIG. 7-25   Rotary joint (a) used in radar application (b).

The TE_{01} mode, whose fields are expressed by equations (7-37), has the characteristic of a very low attenuation when operating at extremely large frequencies. The attenuation for this mode is given by equation (7-38):

$$H_z = H_0 J_0\left(\frac{3.832}{a}\rho\right)e^{-j\beta_g z} \tag{7-37a}$$

$$H_\rho = \frac{j\sqrt{f^2 - [f_c(TE_{01})]^2}}{f_c(TE_{01})} H_0 J_1\left(\frac{3.832}{a}\rho\right)e^{-j\beta_g z} \tag{7-37b}$$

$$H_\phi = 0 \tag{7-37c}$$

$$E_z = 0 \tag{7-37d}$$

$$E_\rho = 0 \tag{7-37e}$$

$$E_\phi = -\frac{j\omega\mu a}{3.832} H_0 J_1\left(\frac{3.832}{a}\rho\right)e^{-j\beta_g z} \tag{7-37f}$$

where a is the radius of the circular waveguide and 3.832 is the first zero of $J_0(x)$:

$$\alpha_{TE_{01}} = \frac{1}{a}\sqrt{\frac{\pi f \epsilon}{\sigma}} \frac{(f_{c(TE_{11})}/f)^2}{\sqrt{1 - (f_{c(TE_{11})}/f)^2}} \frac{Np}{m} \tag{7-38}$$

This equation indicates that the attenuation is primarily governed by the frequency of operation f and the radius of the guide a. Figures 7-26 and 7-27 show the relationship between the attenuation of the guide and its size and frequency of opera-

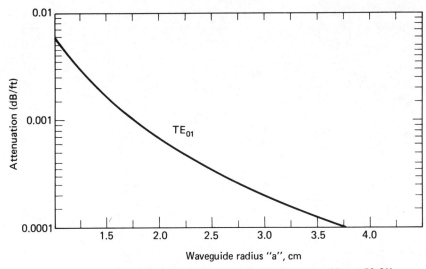

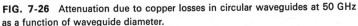

FIG. 7-26 Attenuation due to copper losses in circular waveguides at 50 GHz as a function of waveguide diameter.

tion. Figure 7-27 clearly shows the decrease of attenuation with frequency for the TE_{01} mode.

In an actual communications system where some form of modulation is employed, a range of frequency components will be present. As already illustrated earlier in Fig. 7-12, the group velocity will vary across this frequency spectrum. This results in delay distortion or dispersion. To minimize the delay distortion, the operating frequency should be as large as possible.

From an economic viewpoint the waveguide diameter should be kept small, which means that the cutoff frequency [see equation (7-35)] becomes high. However, delay distortion demands that f/f_c be large. The compromise that seems quite attractive is to employ a waveguide diameter of about 2 in., with the operating frequency at about 50 GHz. This, in turn, brings up the problem of mode conversion.

7-12 MODE CONVERSION

One of the major problems in attempting to utilize the TE_{01} mode is the number of other modes that can propagate, because the TE_{01} mode is not the lowest-frequency or dominant mode. For example, the dominate mode in a 2-in.-diameter waveguide has a cutoff frequency of 3.47 GHz. If an operating frequency of 50 GHz is employed about 200 different modes can propagate. Some of these modes can be coupled to the TE_{01} signal mode through imperfections of the waveguide. Two serious difficulties arise due to this coupling.

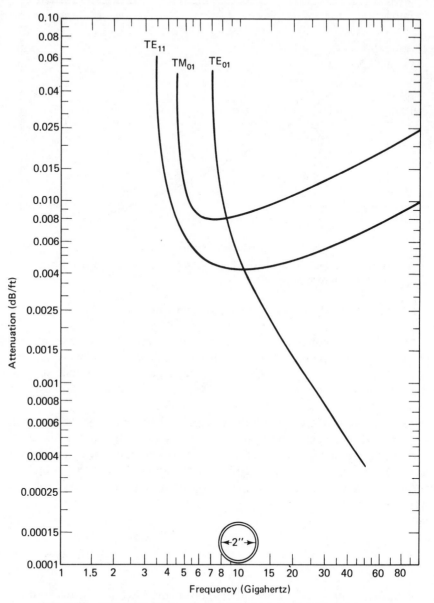

FIG. 7-27 Attenuation due to copper losses in a 2-in.-diameter circular wave-guide. [From E. C. Jordan and K. G. Balmain, *Electromagnetic Waves and Radiating Systems* (Englewood Cliffs, N.J.: Prentice-Hall, Inc., 1968), p. 272, with TE_{01} curve extended.]

First, a conversion at one imperfection can become a reconversion to the signal mode at another imperfection. Since each mode propagates at a different velocity, owing to differing cutoff frequencies, the modes can arrive at the second imperfection at different times. Thus, the second conversion process can result in the TE_{01} mode containing two energy pulses separated in time. This results in a delay distortion of the signal mode. Second, the attenuation constant for these modes is larger than that for the TE_{01}. This can result in excessive losses. For these reasons, mode filters are often used to reduce the possibilities of setting up the unwanted modes.

7-13 MODE FILTERS

Since the TE_{01} mode has only a ϕ electric field component (this is true for all the TE_{0l} modes), many of the other modes which have additional components can be removed or rapidly attenuated by inserting a resistive sheet in the waveguide in such a manner so as not to disturb the TE_{01} mode. For instance, the modes listed in Table 7-2 have, in addition to the ϕ component, the electric field component(s) as noted. Many of these modes can be spuriously excited or coupled from the TE_{01} mode. From Table 7-2 it can be noted that with the exception of the TE_{01} mode (or TE_{0l} modes), all have radial components.

TABLE 7-2 Electric Field Components for Various Modes

TE_{0l}	E_ϕ
TE_{nl} $(n \neq 0)$	E_ϕ and E_ρ
TM_{nl} $(n \neq 0)$	E_ϕ, E_ρ, and E_z
TM_{0l}	E_ρ and E_z

A filter that could attenuate all the modes except the TE_{01} mode could be made by placing radial resistive sheets in the waveguide, as shown in Fig. 7-28. The TE_{01} mode has its electric field normal to these planes and therefore would receive minimum attenuation.

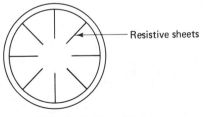

FIG. 7-28 Mode filtering.

Another type of waveguide that also accomplishes attenuation of unused modes of propagation is one constructed of successive copper rings which are insulated from one another. This type of structure provides good conductivity (low resistance) around the ring but very poor conductivity in the longitudinal direction.

To see why this type of guide does not attenuate the TE_{01} mode, we must examine the fields at the boundary of the conductor. Substituting $\rho = a$ into equation (7-37), we obtain

$$H_z = H_0 J_0(3.832)e^{-j\beta_g z}$$
$$H_\rho = 0$$
$$H_\phi = 0$$
$$E_z = 0 \qquad\qquad (7\text{-}39)$$
$$E_\rho = 0$$
$$E_\phi = 0$$

since $J_1(3.832) = 0$.

Since surface current flows in a direction that is both normal to the magnetic field intensity and the surface normal (commonly expressed as $\mathbf{J_s} = \mathbf{\hat{n}} \times \mathbf{H}$, where $\mathbf{\hat{n}}$ is a unit vector normal to the surface and $\times$ is the cross product), the wall current travels in the ϕ direction or around the ring. Therefore, the insulated rings do not disturb the current flow associated with the TE_{01} mode, and it therefore propagates with little attenuation.

When considering the other modes, which have H_ϕ components as well as the H_z component at the boundary, there will be, in addition, a z component of current in the walls. These currents will, however, observe a large resistance between the rings, and will be rapidly attenuated. Thus, these other modes are also attenuated.

In order to simplify the fabrication of the spaced-ring structure, a helical waveguide structure is employed. Typically, for a 2-in. copper helix waveguide at 90 GHz, an attenuation of 2 dB/mi is obtainable for the TE_{01} mode. Figure 7-29 shows a typical helical waveguide.

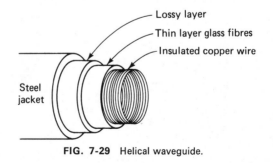

FIG. 7-29 Helical waveguide.

Another structure occasionally used for guiding electromagnetic waves is the dielectric rod, shown in Fig. 7-30. As the equations describing the fields on this type of structure involves both Bessel and Hankel functions, the nature of the propagating waves will be kept to a descriptive basis.

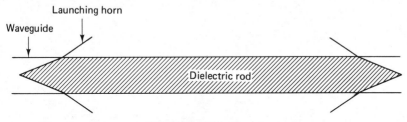

FIG. 7-30 Dielectric rod.

As in the case of rectangular and cylindrical waveguides, both the TE and TM modes can exist. Associated with any one of these modes is a cutoff frequency. If the frequency of operation is kept above the cutoff frequency, the field conducts along the guide with an attenuation due to the dielectric loss of the rod material. However, the fields are not entirely confined within the dielectric rod. Inside the rod a standing wave exists (perpendicular to power flow), and outside the rod, the field strength decays in an exponential fashion. There is no radiation even though some of the field exists outside the guide.

Because of this field existing outside the guide proper, supporting structures cannot be employed. For this reason the dielectric rod can be used for only short distances. If the operating frequency is below the cutoff frequency, energy does escape from the guide. Propagation still occurs, but with a continuous loss due to radiation.

Another mode, not encountered in hollow-pipe guides, called the HE_{11} *mode*, can also exist on such a structure. This mode, which is used in dielectric radiators or dielectric horns for use as antennas, is a "hybrid" or combination TE and TM mode. This mode has no cutoff frequency, although at the lower frequencies it becomes very difficult to launch the wave. The HE_{11} can be launched from the TE_{10} mode in the rectangular waveguide or the TE_{11} mode in the cylindrical waveguide.

7-16 OPTICAL FIBERS

An attractive transmission method in telecommunications is an optical communication system using a single or a multimode fiber waveguide. It consists of a silicate glass or similar material fiber having a diameter in the range 10–400 μm encased in a cladding of slightly smaller dielectric constant. This can result in a very low loss cable with attenuations down to 0.5 dB/km for high-silica fibers. The non-inductive fibers are immune to interference from lightning or electromagnetic sources, to crosstalk, and maintain good electrical isolation between the transmitter and receiver. Figure 7-31 shows a typical optical fiber with its associated relative field strength. With the field in the cladding glass experiencing an exponential decay in intensity in the radial direction, a sufficiently thick cladding will prevent any supports or external objects from interferring with the field structure in the fiber.

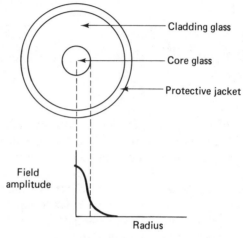

FIG. 7-31 Quartz fiber.

Instead of the traveling wave being bound by conducting surfaces as in the case of metal waveguides, the wave is guided by the two dielectric regions, with most of the energy concentrated within the central core. No conduction currents exist, but displacement currents do, and as a result most modes have a component of both electric and magnetic fields in the direction of propagation.

Consider the transmission behavior from a light-ray approach. If the reader is not familiar with this, refer to Section 8-4. When a light ray inside the core meets the boundary between the core and cladding at an angle θ as shown in Fig. 7-32, total internal reflection occurs provided that

$$\theta < \theta_{\text{cc}}$$

where

$$\theta_{\text{cc}} = \cos^{-1}\sqrt{\frac{\epsilon_{\text{clad}}}{\epsilon_{\text{core}}}} \qquad (7\text{-}40)$$

θ_{cc} is the complementary angle to the critical angle θ_c referred to in Section 8-4; this causes the slight change of equation (8-6) to that of (7-40). Any ray that exceeds this complementary critical angle will propagate into the cladding and be dissipated.

Because only rays that do not exceed this angle can propagate with little loss, the fiber effectively only accepts rays that lie within a certain acceptance cone, as

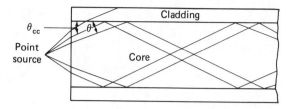

FIG. 7-32 Light rays propagating down an optical fiber.

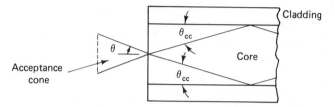

FIG. 7-33 Acceptance cone of a fiber.

illustrated in Fig. 7-33. The term *numerical aperture* of a fiber describes this feature and is defined as the sine of θ, the half-angle of the acceptance cone.

Although ray theory allows for easy visualization of what occurs in a fiber, wave theory gives a more quantitative analysis. If the fiber-core diameter is in the order of 10 μm or less, only single-mode propagation occurs. A short pulse of light transmitted along this type of fiber suffers a small amount of spreading in time, owing to the dependence of dielectric constants on frequency. Since a pulsed signal consists of many frequency components, each frequency experiences a slightly different velocity. This results in a broadening of the received pulse.

Because of the small diameters involved in monomode fibers, axial or lateral alignment is difficult to control when splicing or coupling cables. If, for instance, a lateral offset of one core radius occurs, a loss of about 4 dB results. Fusion splicing techniques can, however, reduce this to well under $\frac{1}{2}$ dB by pulling the fiber back into line. Since monomode fibers with losses of around 0.5 dB/km are being continuously drawn up to 15 km in length, splicing may be less of a problem with fibers than often anticipated. The small diameter of monomode fibers also results in a low numerical aperture, resulting in poor light-gathering efficiency.

An alternative to the monomode fiber is the step-index multimode fiber illustrated in Fig. 7-34(b). It has a much-larger-diameter core, thus making alignment less stringent and resulting in an improvement of numerical apertures. Because many modes can exist in the larger-diameter structure, each with its unique velocity of propagation, pulse dispersion becomes a major problem. A ray that enters the fiber at an angle to the fiber axis travels a much longer distance as it zigzags toward the receiver than an on-axis ray, which travels the shortest distance. This results in pulse broadening and hence the lowering of the maximum permissible pulse rate so as to allow clear distinction between neighboring pulses by the detector. A zero rise time in light intensity at the input results in a nonzero rise time at the output. This mechanism of pulse dispersion is called *modal dispersion*.

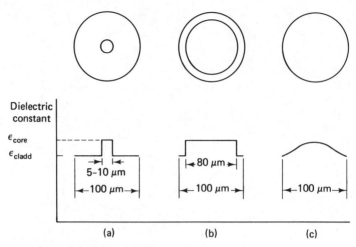

FIG. 7-34 Fiber dielectric profiles: (a) monomode; (b) step-index multimode; (c) graded-index multimode.

As in the monomode case, *material dispersion* also takes place; because of the variation of dielectric constant with wavelength or frequency, this affects the propagation velocity. Any real source has a certain line width (with a corresponding bandwidth), which varies from about 35 nm for LED's and 2 nm for solid-state lasers.

One method of reducing the dispersion difficulty of the step-index fiber is to go to the graded-index multimode fiber shown in Fig. 7-34(c). It has a dielectric profile that is highest at the core center and decreases parabolically until it matches the cladding dielectric at the core-clad interface. Because the rays that traverse the lower dielectric regions travel faster, the off-axis rays, which must travel a greater distance, can reach the output at approximately the same time as the on-axis rays, which experience the highest dielectric. Although it is still a multimode fiber, the graded-index profile minimizes the propagation delay differences between the various modes, resulting in small dispersion. Bit rates of several gigabits per kilometer can be realized.

Cables consisting of several fibers are being manufactured to allow for several communication paths in one sheath and to provide for spares in case a fiber becomes unsuitable for operation. In order to pull such cables through ducts, Kevlar or steel strands are added for tensile strength.

7-17 WAVEGUIDE RESONATORS

When operating above a few gigahertz, the resonant lines discussed in Section 3-9 become impractical, as there is a strong danger that higher modes can be excited. Also, the Q's become unreasonably small. For these reasons, resonant cavities are employed as resonant circuits at the higher frequencies.

The rectangular cavity is formed by enclosing both ends of a rectangular waveguide. As in the case of resonant transmission lines, the length of the cavity is made a multiple number of one-half wavelengths (waveguide wavelengths) long. Many modes can again be excited, depending upon the size of the cavity and frequency of operation. A third subscript is added to the mode designation, indicating the number of half-wave-field intensity variations along the length of the cavity. Thus, the lowest mode for the rectangular cavity would be designated TE_{101}. This mode pattern is shown in Fig. 7-35. In the case of a cavity constructed of copper, the theoretical Q for the TE_{101} mode is given by

$$Q_{TE_{101}} = 10.7\sqrt{f} \qquad (7\text{-}41)$$

which can give Q's of several thousand at microwave frequencies.

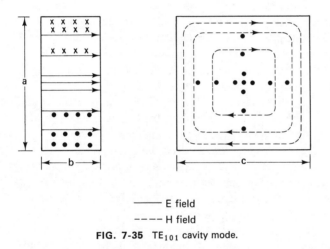

——— E field
- - - - H field

FIG. 7-35 TE_{101} cavity mode.

In practice, however, the introduction of a coupling system and the imperfections in the walls result in a reduced Q. Also the introduction of a dielectric will further reduce the Q due to dielectric losses. Occasionally, walls are silver-plated, to reduce losses in the metallic walls.

In a very similar vein, a cylindrical cavity can be designed by enclosing the ends of a cylindrical waveguide as shown in Fig. 7-36. Here, also, a third subscript is added to denote the number of half-wave-field-intensity variations that occur in the longitudinal direction.

Since the Q's of any resonant cavity vary as the ratio of volume/surface area of a cavity (i.e., Q depends upon the ratio of energy stored to energy lost as indicated in Section 3-9), the Q's of a circular cavity tend to be slightly higher than that of a rectangular cavity. For a cylindrical cavity having the same height/diameter ratio as the height/base ratio of a square-base rectangular cavity, the Q is a little over 8% higher.

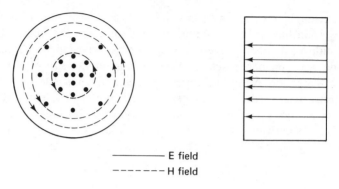

——————— E field

— — — — — — H field

FIG. 7-36 TM_{010} cavity mode.

Also of note is that the Q increases as the mode order increases. The danger of achieving a higher Q by employing a higher-mode cavity is that more than one mode can be excited. This can effectively result in a reduced Q, since power can be expended in setting up this undesired mode. Mode filters can be used to attenuate the unwanted mode, but there is an inherent danger of increasing the losses of the desired mode also, causing a decrease of Q.

Cavities of other shapes are also employed, particularly in the case of active devices, such as the klystron and magnetron. Often, a coaxial type of resonant cavity is used. An example of the latter is the one shown in Fig. 7-37, in which two closely spaced signals receive widely divergent attenuation in a repeater station. It is general practice to retransmit the signal at a frequency different from the received signal in a repeater station, such as a communications satellite or microwave tower, as it is very difficult to prevent some feedback from the transmitting antenna to the receiving antenna resulting in amplifier oscillation.

In this particular system, a loop is used in conjunction with a cylindrical cavity to prevent the transmitted signal (147.00 MHz) from appearing at the receiver input (operated at 146.46 MHz). Two separate antennas are employed, one for the receiver and one for the transmitter. The loop is $1\frac{1}{2}\lambda$ long at a frequency between the desired pass and reject frequencies. The cavity is at series resonance at the pass frequency (146.46 MHz). When the cavity, being series-resonant, is connected to the loop at C, points A and B ($\lambda/4$ from C) observe an "open" when looking toward C. The received signal at A therefore passes around to D to the receiver input at B. The stub at D has little effect, since it appears as a high impedance. An insertion loss of 0.5 to 1 dB is typical. At the reject or transmitted frequency of 147.00 MHz, the cavity being off resonance presents a large reactance at C as a result of its high Q. At point D a stub is inserted to present the same reactance at point D as the cavity presents at point C. Now the input at A sees two signal routes: A via D to B and A via C to B. Both routes offer the same signal

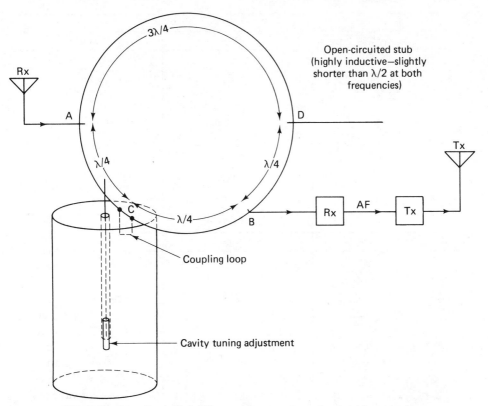

FIG. 7-37 Example of a hybrid loop-cavity resonator system used in a repeater station.

disturbance, but the one A via D has a length of 1λ compared to path via C, which has a length of $\lambda/2$. Therefore, these two signals arrive at B 180° out of phase and cancel. Rejection of the transmitted signal is on the order of 50 to 60 dB.

7-18 WAVEGUIDE COMPONENTS

With some understanding of the fields in a waveguide, of incident and reflected waves, the role of various waveguide components can be generally comprehended. Table 7-3 lists some of the more common waveguide components, with a brief indication of their applications.

TABLE 7-3 Tabulation of Waveguide Components

BENDS		
Physical shape	*Type*	*Application or function*
	H plane	Change direction of propagation.
	E plane	Minimize reflections by making bend several wavelengths long.

TRANSITIONS		
Physical shape	*Type*	*Application or function*
	Twist	Change direction of polarization.
	Rectangular to circular	Convert from rectangular to circular waveguide

JUNCTIONS		
Physical shape	*Type*	*Application or function*
	H-plane tee	Combining or splitting signals.
	E-plane tee	

TABLE 7-3 (Continued)

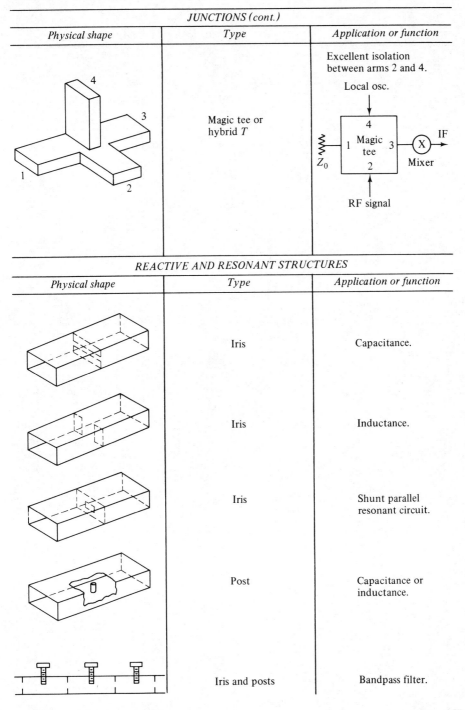

JUNCTIONS (cont.)		
Physical shape	*Type*	*Application or function*
	Magic tee or hybrid *T*	Excellent isolation between arms 2 and 4.

REACTIVE AND RESONANT STRUCTURES		
Physical shape	*Type*	*Application or function*
	Iris	Capacitance.
	Iris	Inductance.
	Iris	Shunt parallel resonant circuit.
	Post	Capacitance or inductance.
	Iris and posts	Bandpass filter.

TABLE 7-3 (Continued)

TUNERS

Physical shape	Type	Application or function
	Slide-screw	Single-stub tuner.
	E-H tuner	Double-stub tuner.

ATTENUATORS

Physical shape	Type	Application or function
	Glass-vane	Variable attenuator.
	Flap	Variable attenuator.
Coaxial line — Coaxial line — TE_{11} mode — Circular waveguide beyond cutoff	Beyond cut-off	Variable attenuator.
TE_{11} Circular waveguide Rotating vane	Rotating vane	Variable attenuator.

TABLE 7-3 (Continued)

FERRITE ISOLATORS		
Physical shape	*Type*	*Application or function*
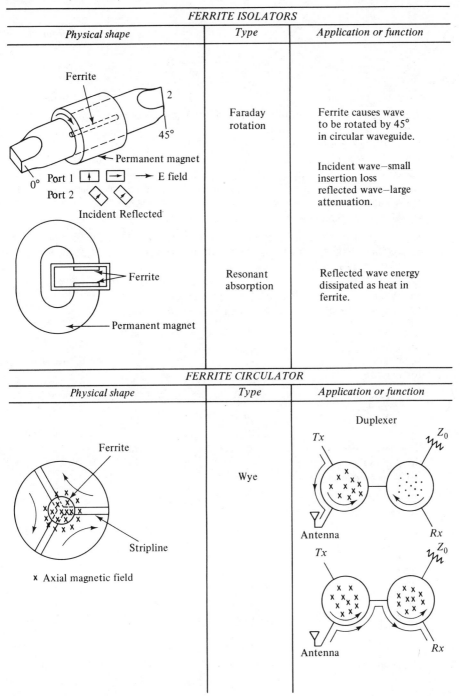	Faraday rotation	Ferrite causes wave to be rotated by 45° in circular waveguide. Incident wave—small insertion loss reflected wave—large attenuation.
	Resonant absorption	Reflected wave energy dissipated as heat in ferrite.

FERRITE CIRCULATOR		
Physical shape	*Type*	*Application or function*
	Wye	

PROBLEMS

7-1.

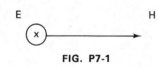

FIG. P7-1

A plane wave traveling in free space (region 1) consists of a horizontal E field and a horizontal H field, as shown.
(a) Indicate the direction of propagation of the wave.
(b) If $E = 2$ V/m after the wave enters a wide lossless medium (region 2) where $\epsilon_r = 4$ and $\mu_r = 2$, calculate H.
(c) Calculate the wavelength in each region if the operating frequency is 100 MHz.
(d) What is the velocity of wave propagation in each region?
(e) Determine the phase constant in each region.

7-2. Corning glass 0120 has a relative dielectric constant of 6.64 at 3 GHz, with a corresponding dissipation factor of 0.0041 (from the 4th edition of *Reference Data for Radio Engineers*, ITT Corp.).
(a) What is the significance of the dissipation factor of a dielectric?
(b) What is the conductivity of Corning glass 0120?
(c) Is this material a good conductor or a good dielectric?

7-3. For conductors: (d) $\sigma \ll \omega\epsilon$
(a) $\mu \gg \omega\epsilon$ (e) $\eta \gg \mu$
(b) $\sigma \gg \omega\epsilon$
(c) $\sigma = \omega\epsilon$

7-4. At a frequency of 1 GHz, a material has the following electrical characteristics:

$$\sigma = 1000 \text{ S/m}$$

$$\epsilon_r = 4$$

This constitutes:
(a) A good conductor.
(b) A good dielectric.
(c) Neither (a) nor (b).

7-5. Match for closest analogy:

Field	Line
(a) μ ____	1. C
(b) ϵ ____	2. R
(c) σ ____	3. L
(d) η ____	4. G
(e) γ ____	5. Z_0
(f) H ____	6. I
	7. V
	8. None

7-6. A plane EM wave in free space impinges normally upon a lossless medium (of infinite thickness) having the following properties:

$$\epsilon_r = 4$$
$$\mu_r = 2$$
$$\epsilon_0 = \frac{1}{36\pi} \times 10^{-9} \text{ F/m}$$
$$\mu_0 = 4\pi \times 10^{-7} \text{ H/m}$$

Obtain the following for a frequency of 1 GHz:
(a) λ in free space.
(b) v in the medium.
(c) λ in the medium.
(d) The intrinsic impedance of the medium.
(e) The VSWR in the free-space region.
(f) The VSWR in the medium.

7-7. A plane wave impinges normally upon a dielectric slab of polystyrene 1 m thick ($\epsilon_r = 2.55$, $\mu_r = 1$). If the frequency of the wave is 100 MHz, what is the electric field reflection coefficient at the two boundary interfaces? (The problem can be solved by using the Smith chart and the transmission-line analogy as shown in the figure.)

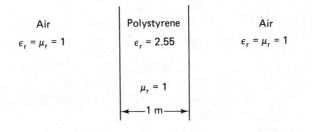

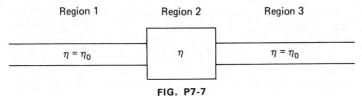

FIG. P7-7

What is the VSWR in:
(a) Region 1?
(b) Region 2?
(c) Region 3?

7-8. Define the following:
(a) Depth of penetration.
(b) Plane wave.
(c) Intrinsic impedance.
(d) Dissipation factor.
(e) Tangential electric field boundary condition.

7-9. Draw an equivalent transmission-line circuit of a lossy dielectric media. Supply all pertinent symbols and notations.

7-10. (a) What is the depth of penetration of a 1-GHz plane wave in a copper sheath?

(b) What is the minimum thickness of Cu required to consider the wave as being totally attenuated?

7-11. An air-filled hollow rectangular metal waveguide has dimensions $a = 1$ in. and $b = \frac{1}{2}$ in.

(a) What is the TE_{10} mode cutoff frequency?

(b) What is the TE_{11} mode cutoff frequency?

7-12. (a) What is the difference between a TEM and a TE mode?

(b) Explain the difference between phase and group velocity as applied to waveguides.

(c) Define the terms "cutoff wavelength" and "dominant mode" as applied to waveguides.

7-13. A rectangular waveguide having inside dimensions of 0.4 in. $\times$ 0.9 in. is propagating in the TE_{10} mode at 10 GHz.

(a) Draw the electric and magnetic field configurations in the waveguide.

(b) Determine the cutoff frequency for this mode.

(c) Determine the free-space wavelength.

(d) Determine the phase velocity of the wave.

(e) Determine the waveguide wavelength.

(f) Determine the velocity of energy propagation of the wave.

(g) Draw a wave launcher that would excite the TE_{10} mode in the waveguide from a coaxial line.

(h) What are the cutoff frequencies for the first two higher-order modes?

7-14. The cutoff frequency of a TE_{10} mode in an air-filled rectangular waveguide is 3 GHz. What would the cutoff frequency be if the same waveguide is filled with a lossless dielectric having an $\epsilon_r = 3.24$?

7-15. Sketch the electric field configurations of the TE_{21} mode.

7-16. (a) Noting in an air-filled rectangular waveguide that $\lambda_g = c/\sqrt{f^2 - f_c^2}$ [see equation (7-28)] and that $\lambda_0 = c/f$ and $\lambda_c = c/f_c$; prove

$$\lambda_0 = \frac{\lambda_g \lambda_c}{\sqrt{\lambda_g^2 + \lambda_c^2}}$$

(b) In a rectangular waveguide having inside dimensions 1 cm $\times$ 2.3 cm the distance between voltage nodes is measured to be 2.1 cm. Using the equation derived in part (a), find the frequency of operation.

7-17. To achieve minimum attenuation in the guide shown in Fig. 7-16, what should the operating frequency be in terms of the cutoff frequency? Assume that the TE_{10} mode is employed.

7-18. (a) Obtain the theoretical Q of a copper air-filled rectangular cavity operating in the TE_{101} mode at 9 GHz.

(b) Suggest a method of coupling a coaxial line to the cavity in part (a).

7-19. (a) The circular waveguide propagating the TE_{01} mode appears to be very promising for future communication systems. Discuss the advantages and some of the difficulties with using this mode of propagation. In particular, consider losses, mode-conversion problems, and techniques of reducing these difficulties.

(b) Show a method of exciting the circular TE_{01} mode from a coaxial line and a mode filter to prevent the TE_{11} mode from propagating.

7-20. For the X-band RG52/U waveguide, obtain the following:

(a) Inside dimensions.

(b) Outside dimensions.

(c) Recommended operating frequency range for the TE_{10} mode.

(d) Attenuation range (in dB/100 ft).

7-21. Describe the advantages and disadvantages of:

(a) Monomode fibers.

(b) Step-index multimode fibers.

(c) Graded-index multimode fibers.

eight

RADIO-WAVE PROPAGATION

8-1 INTRODUCTION

Up to this point we have dealt with structures that guide electro-magnetic waves from one point to another. When transmitting over great distances, it is often more efficient and economical to let the electromagnetic waves propagate through free space, even though the waves tend to spread out in a more-or-less spherical pattern. In regions distant from a transmitting antenna, the power density takes on an inverse relationship with the distance rather than assuming an exponential decay as with the guiding-type structures, resulting in lower overall losses. Although radiation from an antenna, which acts as a matching device between a source and free space, occurs at all frequencies, its relative magnitude is insignificant until the antenna becomes an appreciable portion of a wavelength in size.

In this chapter we shall consider the various mechanisms by which electromagnetic waves propagate between two points. Figure 8-1 pictorally depicts the types of propagation paths that will be considered.

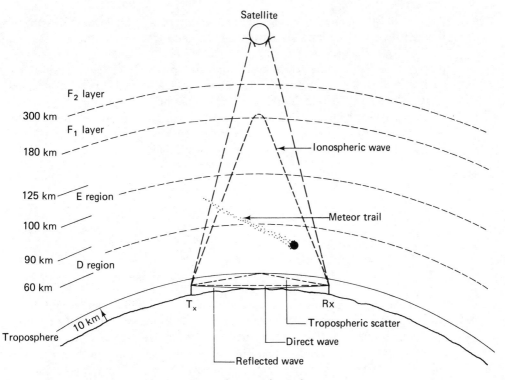

FIG. 8-1 Types of propagation paths.

8-2 POLARIZATION

The plane electromagnetic wave shown in Fig. 7-1 is *linearly polarized*; that is, the direction of the electric field has a constant orientation in space regardless of the time. This particular wave is polarized in the x direction. In general, the polarization of a wave refers to the time-varying behavior of the electric field vector at some fixed point in space.

Often, when considering waves near the earth's surface, we say that a wave is vertically polarized when the electric vector is vertical or lies in a vertical plane. An example of this is the commercial broadcast AM signal (it, however, also has a horizontal component along the direction of propagation, as a result of the loss in the ground). If the wave lies in the horizontal plane, the wave is said to be *horizontally polarized*. An example of this is the TV signal in North America.

The polarization of the wave is initially determined by the launching antenna, but it may change when going through, for instance, an ionized media (e.g., Van Allen radiation belt). For this reason circular or elliptical polarization (where both vertical and horizontal polarization components are present) is employed at times.

Also, artificial satellites may change their polarization relative to the earth's station, so the receiving antenna on the earth should employ elliptical polarization.

The circular polarized wave has its electric field rotate about the direction of propagation so that the wave makes one full rotation for each wavelength it advances. Such a wave can be produced by two linearly polarized antennas positioned at right angles to each other and fed 90° out of phase (the turnstile antenna). If the signals to the two antennas of Fig. 8-2 are of different amplitudes, elliptical polarization results. For an elliptically polarized wave, the axial ratio, the elliptically angle, and the direction of rotation are specified.

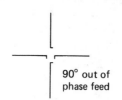

90° out of
phase feed

FIG. 8-2   Antenna producing circular polarization.

According to the IEEE Standards, the wave is said to have right-handed circular polarization when the wave rotates in a clockwise direction when receding from an observer. A right-handed helical antenna (see Table 9-1) transmits or receives right-circular polarization. Thus, if at a fixed position on the z axis the E field rotates in a clockwise direction, as illustrated in Fig. 8-3, the wave is said to be right-handed polarized when traveling in the z direction.

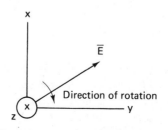

Direction of rotation

FIG. 8-3   Right-circular polarized wave.

Figure 8-4 shows the electric field orientation of the various types of polarization discussed. Random polarization occurs when there is no fixed polarization pattern, as in the case of many noise sources (galatic noise, electric light bulb, etc.).

The polarization of a field radiated from an antenna can be measured by rotating a linearly polarized receiving antenna in the field. As the polarization of the latter is rotated, two maxima and two nulls are observed for a linearly polarized field. The maxima indicates the polarization. (The direction of the nulls can be measured more accurately than the maxima and is usually used by noting that the maxima are 90° from the nulls.)

In an elliptically polarized field there will be no nulls, only minima. The

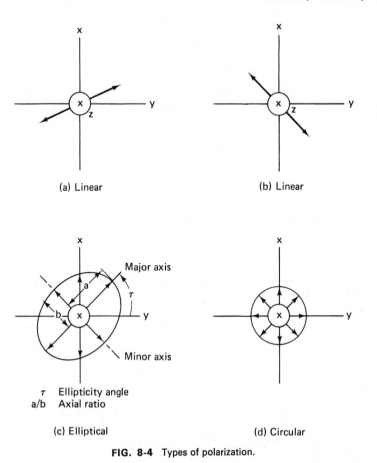

(a) Linear   (b) Linear

(c) Elliptical   (d) Circular

τ Ellipticity angle
a/b   Axial ratio

FIG. 8-4   Types of polarization.

axial ratio is obtained by placing the linearly polarized receiving antenna in the main beam of the antenna under test, and obtaining the ratio of the maximum to minimum readings.

8-3 INVERSE-SQUARE LAW

From the law of conservation of energy, the total power crossing a sphere surrounding an antenna is constant regardless of the sphere's diameter, provided that the medium is lossless and that no additional power is supplied or removed. If we consider an isotropic radiator, which is a fictitious source that radiates uniformly in all directions, the power density is reduced as one moves away from the radiator, since the constant total power radiated is scattered over a larger surface.

Let the total transmitted power be P_t and the radius of the two spheres be R_1 and R_2, as shown in Fig. 8-5. The power density at radius R_1 is $\mathcal{P}_1 = P_t/4\pi R_1^2$ and at radius R_2 is $\mathcal{P}_2 = P_t/4\pi R_2^2$, where $4\pi R^2$ is the surface area of a sphere.

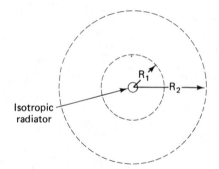

FIG. 8-5 Spherical waves radiating from an isotropic source.

The ratio of the power densities at the two radii is

$$\frac{\mathcal{P}_2}{\mathcal{P}_1} = \left(\frac{R_1}{R_2}\right)^2 \tag{8-1}$$

This expression indicates that the power density is inversely proportional to the square of the distance from the source.

This can be more suitably put into logarithmic terms by comparing the loss in field strength if the distance from the source is doubled. Let $R_2 = 2R_1$; then

$$\frac{\mathcal{P}_2}{\mathcal{P}_1} = \left(\frac{R_1}{2R_1}\right)^2 = \frac{1}{4}$$

Expressing this in decibels,

$$10 \log \frac{\mathcal{P}_2}{\mathcal{P}_1} = 10 \log \frac{1}{4} = 10(-0.602) = -6 \text{ dB}$$

or a 6 dB loss is incurred when the distance from the source is doubled. For example, there would be a 6 dB difference in power density levels at a point 10 km away from the antenna when compared to the density at 5 km. At 20 km the attenuation would be 12 dB when compared to the 5 km position.

At a sufficiently large distance from any antenna, the field as seen over a finite area appears very much like a plane wave. The spherical surface assumes a planar form, and the **E** and **H** fields are perpendicular to one another and contain no radial components. In Fig. 8-6, over the area dA, the wave can be assumed to be a plane wave.

8-4 REFLECTION AND REFRACTION

When dealing with electromagnetic waves reflecting from a surface or undergoing bending when passing through a medium of different dielectric or permeability, it is often more convenient to use the principles developed in the science of geometric optics. The same results can be obtained by employing Maxwell's equations with appropriate boundary conditions, but this requires extensive use of vector calculus.

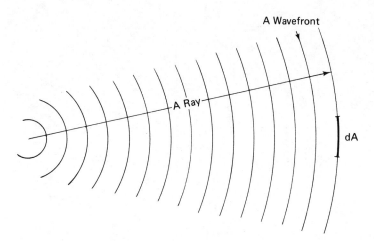

FIG. 8-6 Radiating spherical wave.

Although the latter technique is very useful when considering different wave polarizations, nonuniform dielectric media, and the like, sufficient understanding of
wave behavior can be obtained for our requirements by using geometric optics.
This method is valid as long as the physical dimensions are large compared to the
wavelength of operation and the media involved is not frequency-dependent.

The direction of propagation of a wave is often called a *ray*. The surface
formed by a constant phase on a wave is called a *wavefront*. This nomenclature is
indicated in Fig. 8-6.

When a train of waves strikes a surface of a medium having a different
velocity of propagation, a bent or refracted wave is obtained in the second medium.
Consider the plane incident wave having an angle of incidence θ_i, as shown in
Fig. 8-7. Angles are generally referenced to the normal of the plane in consideration. Assuming that the velocity of the wave in region 2 is lower than that in region
1, while the plane wave in region 1 moves from A to A' in time t, the wave in
region 2 advances from B to B' during the same time interval. Thus, the distances
AA' and BB' can be expressed as v_1t and v_2t, respectively.

By considering the two right triangles $AA'B$ and $BB'A'$, we obtain

$$\sin \theta_i = \frac{v_1 t}{A'B}$$

and

$$\sin \theta_t = \frac{v_2 t}{A'B}$$

Hence,

$$\frac{\sin \theta_i}{\sin \theta_t} = \frac{v_1}{v_2} \qquad (8\text{-}2)$$

This equation is often expressed in terms of the index of refraction of the media,
which is defined as the ratio of the velocity of the wave in a vacuum to the velocity

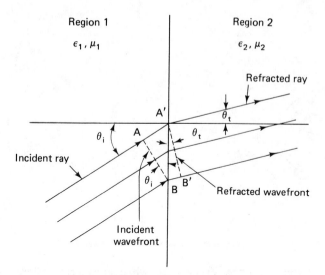

FIG. 8-7 Wave refraction.

of the wave in the medium under consideration. Thus,

$$n \text{ (index of refraction)} = \frac{c}{v} \tag{8-3a}$$

$$= \frac{\sqrt{\mu\epsilon}}{\sqrt{\mu_0\epsilon_0}} \tag{8-3b}$$

$$= \sqrt{\mu_r\epsilon_r} \tag{8-3c}$$

Equation (8-2) can now be rewritten in terms of the relative permeability and dielectric constants. This equation is known as *Snell's law*:

$$\frac{\sin \theta_i}{\sin \theta_t} = \frac{n_2}{n_1} = \sqrt{\frac{\mu_{r_2}\epsilon_{r_2}}{\mu_{r_1}\epsilon_{r_1}}} \tag{8-4}$$

Since the relative permeability constant of most media is unity, equation (8-4) can be expressed as

$$\frac{\sin \theta_i}{\sin \theta_t} = \sqrt{\frac{\epsilon_{r_2}}{\epsilon_{r_1}}} \tag{8-5}$$

If a wave therefore travels from a region of high dielectric constant to one of a lower dielectric constant ($\epsilon_2 < \epsilon_1$), $\sin \theta_t$ must be greater than $\sin \theta_i$ or $\theta_t > \theta_i$, and the ray will be bent away from the normal to the plane, as shown in Fig. 8-8.

Examining equation (8-5), we can note that if $\sin \theta_i$ is made equal to $\sqrt{\epsilon_{r_2}/\epsilon_{r_1}}$, $\sin \theta_t$ becomes equal to unity and θ_t is forced to $\pi/2$ or 90°. This means that no wave phenomena is present in the lower dielectric material. This particular angle of incidence is called the *critical angle θ_c* and is given by

$$\theta_c = \sin^{-1} \sqrt{\frac{\epsilon_{r_2}}{\epsilon_{r_1}}} \tag{8-6a}$$

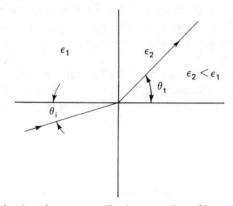

FIG. 8-8 Refraction of a wave traveling into a medium of lower dielectric constant.

or more generally by

$$\theta_c = \sin^{-1} \frac{n_2}{n_1} \qquad (8\text{-}6b)$$

If the angle of incidence is greater or equal to the critical angle, a wave may propagate along the dielectric boundary with no loss of energy, even though no metal boundaries are present. This concept is employed in the design of transmission media for laser beams as described in Section 7-16.

When a wave encounters a discontinuity, part of the wave may also be reflected. Since the velocity of the wave does not change after reflecting from the boundary, the angle of reflection, θ_r, is equal to the angle of incidence, θ_i. Figure 8-9 depicts both a reflected and refracted ray.

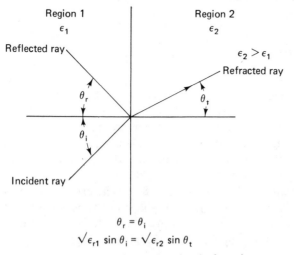

FIG. 8-9 Single incident, reflected and refracted rays.

Reflection techniques are used in many antennas that use reflectors as part of their construction. A good example of this is the parabolic reflector used in conjunction with an antenna or the parabolic reflector used in the headlight of an automobile.

A simple reflector used in periscope antenna systems where a beam is, for instance, radiated from the bottom of a tower to a reflector at the top, is shown in Fig. 8-10. This system is usually less expensive than a waveguide run when the distance to the top of the tower exceeds 100 m.

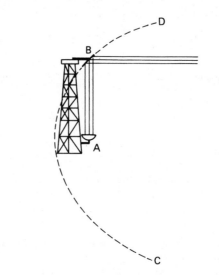

FIG. 8-10 Periscope reflector. The transmitting antenna (A) can be considered as the focus of the parabolic curve *CD*, with the reflector (B) located on the parabolic locus as shown.

8-5 HUYGENS' PRINCIPLE

Huygens' principle is a geometrical method for obtaining the field pattern of a wave, once the field pattern is known at some earlier instant. According to this principle, every point of a given wavefront may be considered as a new radiating source, radiating as an isotropic or point source. The new wavefront, at further points in space, is found by constructing a surface tangent to the secondary wavelets produced by the isotropic sources.

This principle is illustrated in Fig. 8-11. The original wavefront is the spherical surface *AA'*. At a time *t* later, the shape of the new wavefront is obtained by constructing numerous circles of radius $r = vt$ along the original wavefront. The new wavefront is traced by following the surface tangents.

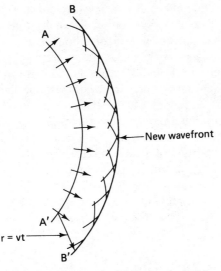

FIG. 8-11 Illustration of Huygens' principle.

8-6   DIFFRACTION

If an absorbing sheet is placed between a radiating source and some observation plane, as shown in Fig. 8-12, one might conclude that no field would exist in the shadow region. In actuality, a field is observed in this region.

Huygens' principle is very helpful in understanding the main features

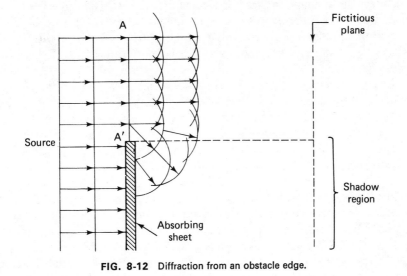

FIG. 8-12 Diffraction from an obstacle edge.

observed in this diffraction effect. If a plane wave encounters the absorbing sheet as shown, every point on the wavefront AA' can be considered as a source of secondary wavelets which spread out in all directions. Thus, some energy will propagate into the shadow region as shown. A little later we will more closely observe the field intensity in such a region, but suffice it to say, at this time, that the field in the shadow region varies in magnitude through several minima and maxima as one moves in the vertical direction. The mathematical analysis is quite involved, finding the resultant of an infinite number of infinitesimal wavelets.

The term "diffraction" is employed in situations where the resultant field is produced by a limited portion of a wave surface. By cutting off a fractional part of a wavefront, the diffraction effect can be noticed.

In the next several sections, we shall consider with the help of the foregoing fundamental concepts, the nature of various categories of radio wave propagation. We will tend to progress through these on a frequency basis, from the VLF spectrum to the EHF's.

8-7   GROUND WAVE

To describe the radiated fields from a vertical antenna over a plane earth having a finite conductivity and operating in the lower-frequency range, we will employ the solutions first developed by Sommerfeld and later refined by Norton. These men showed that the radiated fields form a ground wave that can be divided into three subcomponents: the direct wave, the terrestrially reflected wave (combined to form the space wave), and the surface wave.

The *space wave* is made up from the spherical wave originating from the dipole under the assumption that it is located in free space (the *direct wave*) and a reflected wave. The *reflected wave* is the resultant of a spherical wave originating also from the dipole, but undergoing a reflection from the earth, with the constraint that this incident wave observes a reflection coefficient as if the incident wave was plane.

Since, in general, the incident wave is not plane, particularly if the antenna is close to the earth, the total reflected wave must contain more than that given by the reflected wave just described. These additional terms required make up the *surface wave*. Hence, if an antenna is located at a great distance from the earth, the incident wave at the earth is essentially a plane wave and the surface wave becomes nonexistent. In this case the ground wave is made up entirely by the space wave.

For the vertical antenna near the earth's surface, therefore, both the surface wave and space wave are present. The space wave tends to predominate at large vertical distances from the earth, whereas the surface wave is the main contributor near the surface. The surface wave that is guided along the earth's surface supplies energy to the ground due to the finite conductivity of the earth, and therefore is attenuated. As well as increasing with the resistivity of the earth, this attenuation increases with distance from the antenna and with the frequency of operation. At frequencies around and below the broadcast band, the earth appears resistive.

Ignoring the attenuation of the surface wave for the moment, a typical radiation pattern (the field intensity at a constant radius from an antenna) of a vertical antenna located at the surface of the earth and depicting both the space and surface wave is sketched in Fig. 8-13. Observing the figure we see that space wave varies drastically near the earth when the earth's conductivity is taken into account. For the perfectly conducting earth, the space wave at low grazing angles is at a maximum.

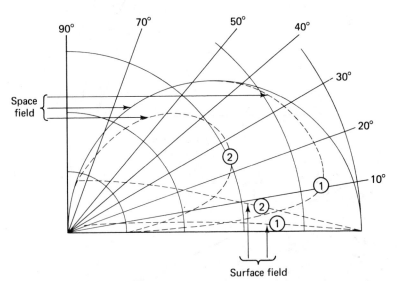

FIG. 8-13 Space and unattenuated surface waves of a vertical dipole located at the earth's surface. Patterns (1) represent conditions over high-conductivity earth in the low broadcast spectrum. Patterns (2) represent conditions over low-conductivity earth in the HF spectrum. ———————— represents field pattern for a perfectly conducting earth. [From E. C. Jordan and K. G. Balmain, *Electromagnetic Waves and Radiating Systems* (Englewood Cliffs, N.J.: Prentice-Hall, Inc., 1968), p. 641.]

Figure 8-14 shows typical field-strength intensities of the surface wave as a function of distance from a vertical antenna located at the surface of the earth. These plots clearly indicate that the higher-frequency surface waves experience greater attenuations and that, in all cases, more attenuation occurs as the distance from the antenna is increased. The inverse distance line ignores all losses but takes into account the inherent spreading effect of the waves as they propagate away from the source.

For horizontally polarized waves, the attenuation factor of the surface wave is much greater than that obtained at the same frequency for a vertically polarized wave. For this reason, horizontal polarization is not employed for antennas near

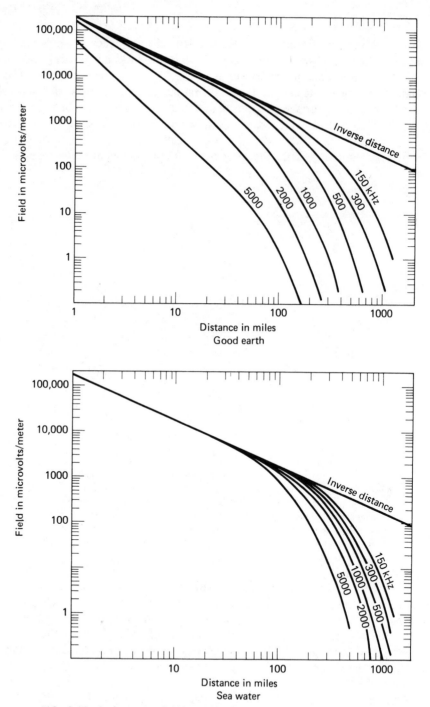

FIG. 8-14 Surface-wave field intensities from a vertical antenna at the earth's surface at and near the broadcast band. (Courtesy International Telephone & Telegraph Corporation. From *Reference Data for Radio Engineers*, 4th ed., pp. 714–715.)

the earth's surface operating at low and medium frequencies. At high and very high frequencies, both polarizations result in high attenuation of the surface wave. In these frequency ranges, elevated antennas are used which propagate via the space wave.

In the VLF range (3–30 kHz), propagation is chiefly by means of the surface wave. Typical attenuation rates of 3 dB/km for transoceanic propagation is experienced around 20 kHz, with land paths being about 3 dB higher. The VLF band requires enormous antennas for efficient radiation and, owing to the rather small frequency spectrum available in this band, are assigned to very few agencies. The radio station WWVL, operated by the National Bureau of Standards in the United States at Fort Collins, Colorado, operates at a carrier frequency of 20 kHz, which can generally be received throughout the world. It is used as a frequency and time standard.

The space wave is nonexistent near the earth's surface when antennas operate at low frequencies near this surface, because the direct and reflected waves of the space wave cancel. The reflection coefficient of the earth at these low grazing angles can be approximated by -1. At increased frequencies, even though the antennas are located near the earth, the space wave is no longer zero as the path-length differences of the direct and reflected wave becomes significant in terms of the smaller operating wavelengths. Thus, the waves no longer arrive at the receiving antenna 180° out of phase and a net signal is received. At the same time the surface wave becomes less significant, so that in and beyond the VHF band, it can be ignored.

For a half-wavelength transmitting antenna carrying a current of I amperes and located a distance of h_1 above the earth, the field strength at a receiving antenna a distance d away and h_2 above the earth is given by equation (8-7) when operating in the VHF band. Referring to Fig. 8-15, this equation is valid under the following approximations:

1. The surface wave is negligible when compared to the space wave.

2. The angle between the incident rays and the earth's surface, ψ_2, is small.

3. The path-length differences (α) between the direct and reflected wave is small, so sin α can be approximated by α.

$$|E| = \frac{240\pi I h_1 h_2}{\lambda d^2} \quad \text{V/m} \tag{8-7}$$

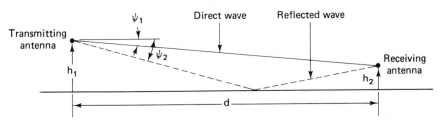

FIG. 8-15 Direct and earth-reflected wave.

The terms h_1, h_2, and d are given in identical units, and λ in meters. This equation is a good approximation for the commonly used FM and TV channels.

8-8 IONOSPHERIC PROPAGATION

In the high-frequency (3–30 MHz) or HF region, waves can also be propagated over long distances by causing the path to refract from the ionosphere. This path is uncontrollable and varies with the time of day, season, and solar activity.

The ionosphere consists of various ionized layers from about 60 to 400 km above the earth's surface. This ionization is caused chiefly by the ultraviolet radiation from the sun, which is much more prominent during the day than in the evenings. In addition, the corpuscular radiation shot out by the sun and arriving at the earth a few days later causes some ionization. The reason for the various layers is that the several gases constituting the atmosphere have different susceptibilities to the ionizing radiation and therefore produce ionization at different altitudes.

During the daytime the ionized regions form four main layers, called the D, E, F_1, and F_2 *layers*, as shown in Fig. 8-16. The D and E layers disappear at night.

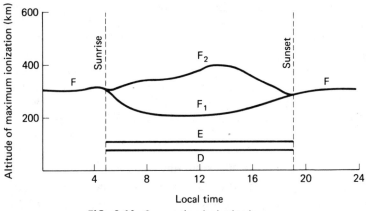

FIG. 8-16 Summertime ionization layers.

In these lower regions the attenuation for the lower frequencies is quite large, resulting in a high absorption of energy of the broadcast signal space wave in the daytime. The F_1 and F_2 layers (which merge into a single F layer at night) are the main layers of interest for ionospheric propagation in the HF band.

In addition to the layers shown in Fig. 8-16, other irregular variations also occur as a result of particle radiation from the sun, resulting in erratic radio-wave propagation. These types of storms can last several days, but communication can still be maintained by lowering the operating frequency.

Excessive erruptions from the sun can cause heavy ionization in the lower regions which causes complete blackout for frequencies above 1 MHz. Signals below this frequency can, however, experience an intensification of the reflected signal. These sudden ionospheric disturbances, sometimes called the *Dellinger effect*, can last for 1 hour or more. Longer-lasting disturbances can also occur, but it usually is still possible to maintain communication. To understand the mechanism of refraction in the ionosphere, we may employ the results developed in many textbooks as they relate to wave propagation through an ionized media. The expressions for both the conductivity and dielectric constants of an ionized gas are very much dependent upon the frequency of operation and the electron density. Although the conductivity expression affects the attenuation of the radio wave, at the frequency where refraction in the F layers becomes of practical importance, the attenuation constant is small and can be neglected for a first approximation. The relative dielectric constant, however, decreases sufficiently as the wave enters the F layer so as to cause the wave to have a noticeable bent. In the frequencies of interest, particularly in the upper portion of the HF band, the relative dielectric constant of the ionized region can be approximated by

$$\epsilon_r = 1 - \frac{Nq^2}{\omega^2 \epsilon_0 m} \tag{8-8}$$

where q = charge of an electron (1.59×10^{-19} C)
   m = mass of an electron (9.1×10^{-31} kg)
   N = electron density

We can note from equation (8-8) that the relative dielectric of an ionized medium is less than that in free space. Substituting equation (8-8) into (8-3c), the index of refraction in an ionized medium becomes

$$n = \sqrt{1 - \frac{Nq^2}{\omega^2 \epsilon_0 m}} \tag{8-9}$$

Since the index of refraction is proportional to the square root of the relative dielectric constant, an electromagnetic wave entering a region such as the ionosphere, where the electron density is increasing, will experience a reduced refractive index or dielectric constant and will be bent away from the normal. The reader should refer back to Fig. 8-8 for an earlier discussion on this phenomenon. As a result, the ray may curve back out of the ionized region, as sketched in Fig. 8-17. This can result in what is called a *sky wave*. The angle of incidence is quite critical since if the receiving antenna is too close to the transmitting antenna, the wave may not bend enough to be received. The area not covered is called the *skip distance*, which is shown in Fig. 8-18.

From equation (8-9) it can be seen that the frequency also is critical. The higher the frequency, the closer the index of refraction approaches unity, and therefore less bending will occur over a similar distance. Ionospheric refraction can generally be employed for frequencies less than 30 MHz and distances greater than 100 miles, although the exact upper limit varies with time of day, the season, and other factors.

The maximum usable frequencies (MUF) for single-hop transmission at

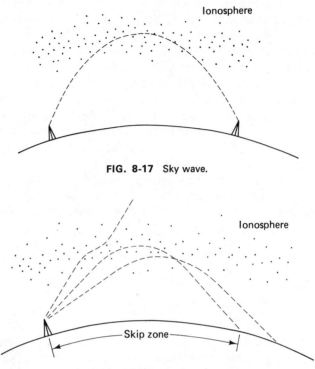

FIG. 8-17 Sky wave.

FIG. 8-18 HF ionospheric propagation.

various distances throughout the day are published by the National Bureau of Standards in the United States. A typical chart is shown in Fig. 8-19. Since the attenuation does decrease with frequency, it is desirable to use as high a frequency as possible. For a more detailed description of ionospheric propagation, one can refer to Chapter 15 of *Communications System Engineering Handbook* by D. Hamsher (New York: McGraw-Hill Book Company, 1967).

Fading and distortion are commonly encountered due to a multipath phenomenon. These multipaths can be formed by a different number of hops between the earth and the F_2 ionized layer (see Fig. 8-20) or between different ionized layers. As a result, the various frequency components in a signal can undergo different delays, resulting in noticeable distortion. Also, phase cancellation can result in part or in whole, resulting in fading. Fading rates result from fluctuation in ionospheric height or absorption.

Transmission is also possible by bouncing VHF signals from the ionized trails of meteors that exist at altitudes of 80 to 120 km. Because the ionized meteor trails last for only short periods of time (usually less than a few seconds), communication is intermittent, and therefore the information is usually sent in bursts. The meteor burst communication system is particularly suited to long-range (up to 1000 mi) low-data-acquisition applications. The rate of incidence of meteor

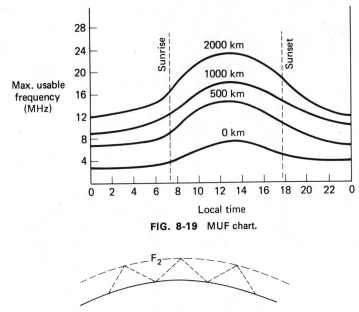

FIG. 8-19 MUF chart.

FIG. 8-20 Propagation by a number of hops.

activity is dependent upon the time of year and the time of day. Maximum yearly occurrences occur in August, with a minimum in February. Because of the movement of the earth around the sun, the morning hours are more active, as the meteors on the evening side must overtake the forward motion of the earth. Typical transmitter powers of 20–2000 W are required.

8-9 LINE-OF-SIGHT PROPAGATION

To obtain more information channels for the increased demands by the public (TV, data communications, voice communications, etc.) bands are continuously being opened up in the higher-frequency ranges. Microwave frequencies are commonly employed, with the chief restriction on terrestrial systems of having a limited range of about 30–40 mi per link, depending upon the topology of the terrain. Propagation takes place in the lower-atmosphere (troposphere) region, where the region is much more stable than, for instance, the ionosphere. Usually less than 10 W of transmitted power is required.

In the troposphere region, the temperature, pressure, and water vapor cause variations in the dielectric constant. This results in changes of the index of refraction, which under normal conditions causes downward bending of the microwave beam in the vertical plane (Fig. 8-21). The amount of bending increases with wavelength, with light beams traveling in straighter lines than a microwave beam.

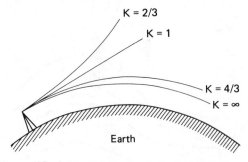

FIG. 8-21 Ray bending in the troposphere.

Two techniques are often employed when plotting profiles of the microwave beam and the earth when checking out microwave paths. One method employs a straight microwave beam, with the curvature of the earth changed so as to compensate for the beam bending. The change in the earth's radius is denoted by the parameter K, where

$$K = \frac{\text{effective earth radius}}{\text{actual earth radius}} \tag{8-10}$$

This method is illustrated in Fig. 8-22. On a compensated earth radius, a typical earth profile for North America is the illustrated $K = \frac{4}{3}$ effective earth radius.

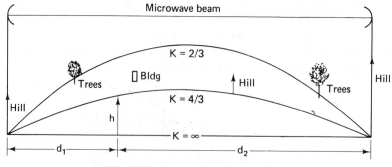

FIG. 8-22 Straight-line microwave beam.

The second method uses a flat earth with the microwave beam having a curvature of $K \times$ the actual earth's radius. This method is illustrated in Fig. 8-23. The profile of the earth's terrain is also superimposed on this sketch. The value most frequently used, $K = \frac{4}{3}$, is referred to as the K factor for the "standard atmosphere." K often increases around sunset and a few hours afterward from the more stable period during daylight.

In general, the "earth bulge," as indicated in Fig. 8-22, is given by

$$h = \frac{d_1 d_2}{1.5K} \qquad \text{ft} \tag{8-11}$$

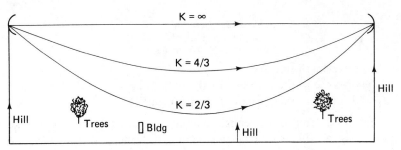

FIG. 8-23   Compensated microwave beams on a flat earth.

(d_1 and d_2 in miles), which results for the standard atmosphere ($K = \frac{4}{3}$) in an h value of

$$h = \frac{d_1 d_2}{2} \qquad \text{ft} \tag{8-12}$$

As the earth and various obstacles may be close to the direct beam, reflection and refraction may also occur. The effect of these depends upon the Fresnel distance, the distance the direct ray is from the earth or obstacle.

Consider two rays propagating from a transmitter (Tx), both reaching the receiver (Rx) as shown in Fig. 8-24. When the path length of the reflected ray is one-half wavelength longer than the direct ray, the reflected wave will arrive in phase with the direct ray if there is 180° phase reversal on the gound reflection.

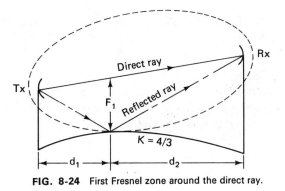

FIG. 8-24   First Fresnel zone around the direct ray.

(This is valid for small grazing angles.) The path clearance F_1, as noted in Fig. 8-24, is called the *first Fresnel zone radius*. The first Fresnel zone is the area within which the path lengths differ by one-half wavelength or less. For reflection coefficients of -1, the received signal will be twice as strong as the direct signal, hence the receiver will see twice the electric field intensity that would occur if the two antennas were in free space, remote from the earth and its effects. Since the power received is proportional to the square of the electric field intensity, this results in a gain of 6 dB over the free-space value. The radius of the first Fresnel zone, F_1, is given by

$$F = 131.5\sqrt{\lambda \frac{d_1 d_2}{d_1 + d_2}} \qquad \text{ft} \qquad (8\text{-}13)$$

where λ = wavelength (meters)

d_1 = distance to the near end of the path (miles)

d_2 = distance to the far end of the path (miles)

If the path difference is extended to a full wavelength by raising the antennas, the resultant signal will be zero, again assuming the reflection coefficient to be -1. This radius is called the *second Fresnel zone radius*, F_2. In general, the Fresnel zones are numbered from the center outward, as shown in Fig. 8-25.

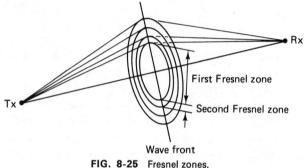

FIG. 8-25 Fresnel zones.

The radius of any Fresnel zone can be expressed in terms of the first Fresnel zone by the equation

$$F_n = \sqrt{n}\, F_1 \qquad (8\text{-}14)$$

where n represents the nth Fresnel zone.

We can now discuss the effects of path clearance on three theoretical types of microwave paths: plane earth ($\Gamma = -1$), smooth sphere ($\Gamma = -1$), and knife-edge diffraction ($\Gamma = 0$). Figure 8-26 indicates the three types of terrain that will be considered.

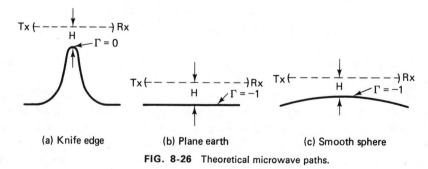

(a) Knife edge   (b) Plane earth   (c) Smooth sphere

FIG. 8-26 Theoretical microwave paths.

To aid us in the discussion, we will refer to Fig. 8-27, which shows the effect of path clearance on radio-wave propagation for the various types of terrain. The inherent loss due to spreading, or the free space loss, is not taken into account. The free space loss must be added to the loss given in the figure. At positive clearances, for the plane-earth or smooth-sphere case, if the clearance passes through any even Fresnel zone, a minimum occurs. This is due to the direct and reflected waves arriving out of phase at the receiver. On the other hand, if the clearance passes through an odd Fresnel zone, a maximum occurs. Thus, as a receiving antenna is raised above the earth, a successive number of maxima and minima field strengths will be observed, with the degree of variation depending upon the magnitude of the reflected wave. The location of the maxima are of a great deal of interest when determining the height of a microwave antenna. It is also possible to determine the location of obstructions in the microwave path if the received signal strength relative to the free-space value is noted as the vertical position of the receiving antenna is varied. This is repeated for a few different transmitting antenna heights, depending upon the number of obstacles present.

For negative clearances, a signal of decreasing amplitude is also obtained due to the diffraction effect (refer to Section 8-6). It should also be noted that a small gain over the free-space value can occur for some positive clearances. Thus, an obstacle may actually enhance a signal, rather than reduce it. In the real-earth case, the received signal for negative clearances will lie somewhere between the knife-edge diffraction and smooth-sphere diffraction curves (see shaded region) as the actual earth's reflection coefficient is somewhat less than 1.

Well-wooded terrain rarely results in any reflected signal and the signal strength therefore remains within a few decibels of free space. Partially wooded terrain, with a mixture of bare fields and trees, can cause severe loss in signal strength due to strong reflected signals. Lakes, rivers, and so on, also cause severe reflections and are avoided if possible along a microwave path. Losses of signal strength exceeding 25 dB are possible. Deserts and large cultivated fields present similar even zones, fading up to 20 dB.

At the higher microwave frequencies, rain also tends to increase the attenuation of the signal, owing to the absorption and scattering of the beam. It is serious only at frequencies above 10 GHz, where the rain drops become an appreciable portion of a wavelength in size. At the same time, microwave signals can also be used to detect rain and hailstorms and are frequently used in aircraft to avoid such disturbances, In these cases the returned signal is monitored as it reflects from the rain or hail.

In some parts of the world, as in the tropical areas and over large bodies of water, where temperature inversions takes place, ducting can occur due to the abnormal variations in the refractive index. Under these conditions, an inverse beam bending can occur or a wave can be trapped near the earth's surface or in some elevated duct.

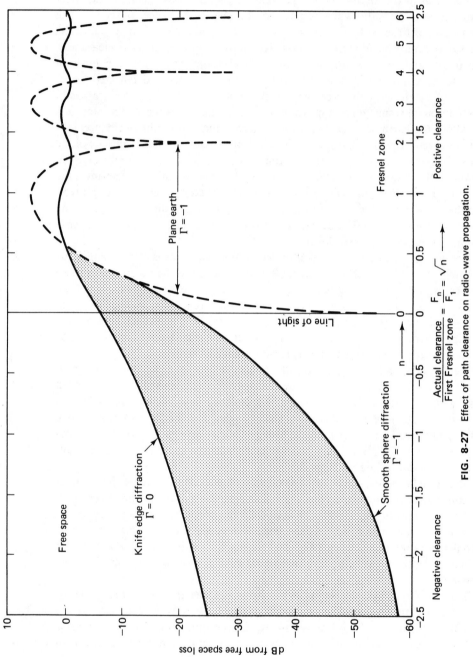

FIG. 8-27 Effect of path clearance on radio-wave propagation.

Tropospheric scatter communication systems, operating in the UHF and SHF bands, are presently being used in areas where sites are inaccessible, or in island hopping, where line-of-sight circuits are extremely expensive. Hops from 70 to 600 mi are made in systems that span across the entire northern region of Canada and down into the United States. Excellent reliability and good information capacity is achieved by using relatively high transmitted power, FM, and highly sensitive receiving apparatus. Large directional antennas are employed resulting in high site-development costs.

The scattering occurs in the tropospheric regions, where there are "blobs" of atmosphere whose refractive index is different from that of the surrounding atmosphere. Partial reflection from the gravitational stratified regions of the atmosphere may also cause the return of the beam toward the earth's surface. Since this scattering occurs in all directions, the received signals are very small. For this reason, high-power transmitters are required, together with high-gain antennas. Frequencies from 100 MHz to 10 GHz can be used. A "tropo" link operated by Alberta Government Telephones, for example, spans a distance of 90 mi and has a capacity of 120 voice channels. The transmitter puts out 5 kW at a frequency of 5 GHz.

8-11 SATELLITE REPEATERS

To obtain reliable communications links over large distances and to achieve worldwide television, communications satellites are being placed in orbit around the earth. Many countries have joined efforts by forming the International Telecommunications Satellite Organization (INTELSAT). The Canadian Corporation (TELESAT) and the American Corporation (COMSAT) form part of the membership.

To maintain continuous communication services between earth stations by satellites, satellites are generally placed in geostationary orbits whereby each satellite is synchronized with the earth's rotation. With three such satellites equally spaced and placed 35,800 km above the equator, the entire globe can be covered. Because of the large slant heights from the earth stations to the satellite (41,200 km for 5° elevation), large nonsteerable earth station antennas are required to obtain the necessary sensitivities. The USSR Molnya series is an exception to this, whereby polar orbits are followed, resulting in improved transmission characteristics to the far north.

INTELSAT operates a global system with satellites placed over the Atlantic, Pacific, and Indian oceans. The INTELSAT IV-A satellite has a height of 5.9 m, weighs 790 kg, and handles 6000 simultaneous telephone circuits (full duplex). It has a lifetime in orbit of seven years, resulting in an investment cost of about $1100 per circuit-year.

Most satellites employ standardized FM techniques, although a few are going digital. The lower-frequency band allocated by CCIR for satellite communications is:

5.925–6.425 GHz   (up path)

3.700–4.200 GHz   (down path)

Because these bands are shared with the terrestrial microwave links and thus run the risk of mutual interference, the power radiated by the satellite must be limited. In the future, higher frequencies will be employed to reduce this problem.

Each satellite acts as a repeater, amplifying and retransmitting the received signals. By using several carriers simultaneously, use can be made of a satellite by several earth stations. This is called *frequency division multiple access* (FDMA). With this scheme, voice channels are frequency-modulated and frequency-division-multiplexed (FM–FDM) onto a preassigned carrier. Each earth station receives all the carriers but selects only the desired channels.

The use of *time division multiple access* (TDMA), whereby each user obtains access to the satellite for short instances of time, has the advantage of being very flexible compared to FDMA. This is because the later requires extensive modification to the receiving and transmitting earth station's equipment if frequencies are reallocated. TDMA causes each station in turn to transmit a burst, which occupies the complete bandwidth of the satellite amplifier; therefore, to reallocate users, only the timing and spacing of bursts need to be altered, as only one carrier is transmitted at any time.

The transmitter in a typical earth station produces a 12-kW signal, which is fed to a parabolic antenna about 30 m in diameter. After traveling about 40,000 km through space, the signal strength at the satellite is in the order of picowatts (7×10^{-12} W). The amplifier translates this frequency to the 4-GHz band and boosts it up to a few watts and beams it back to the ground. At the earth station the 30-m dish antenna collects the few picowatts of energy and feeds it to a low-noise receiver to be amplified.

TELESAT, established by the telecommunications common carriers of Canada and the Canadian government, operates an independent domestic series of satellites called ANIK. ANIK I, II, and III are positioned at 114°, 109°, and 104° West longitude, respectively. Each ANIK satellite covers all of Canada with 12 high-capacity microwave channels, each having 36 MHz of bandwidth, spaced by a 4-MH guardband. Each channel is capable of relaying one color TV program or up to 960 multiplexed voice conversations using a single carrier. Each channel will provide not less than 33 dBW EIRP (effective isotropic radiated power; refer to Section 9-11) throughout Canada. Figure 8-28 shows the EIRP contours in dBW (reference of 1 W). The TWT output power from the satellite can generate 5 W (7 dBW) of power.

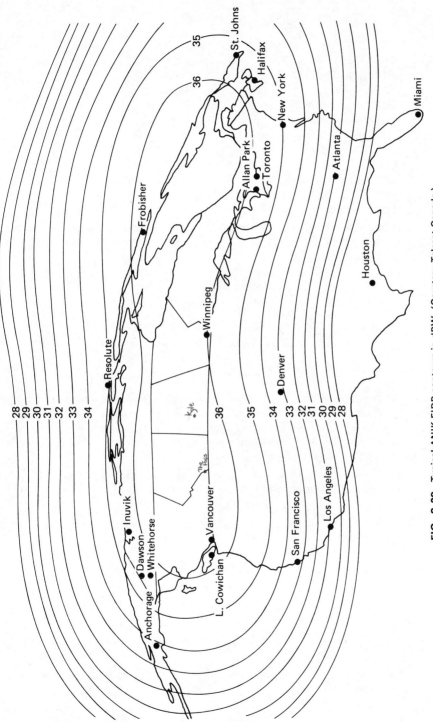

FIG. 8-28 Typical ANIK EIRP contours in dBW. (Courtesy Telesat Canada.)

PROBLEMS

8-1. Define the following terms:
  (a) Polarization.
  (b) Plane wave.
  (c) Reflected wave.
  (d) Refracted wave.
  (e) Diffracted wave.
  (f) Critical angle.
  (g) Surface wave.
  (h) Ground wave.
  (i) Fresnel zone.
  (j) Skip zone.
  (k) Tropospheric scatter.
  (l) Knife edge.

8-2. A spherical electromagnetic wave has a power density of 1 mW/m² at a distance of 100 km from the source. What is the power density 400 km from the source if the wavefront is regarded as spherical? What is the attenuation in decibels between these two points?

8-3. A plane wave in free space impinges upon a plane dielectric surface at an angle of incidence of 30°. If the relative dielectric constant of the material is $\epsilon_r = 3$, what is the angle of refraction?

8-4.

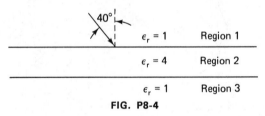

FIG. P8-4

A plane wave in free space is incident upon a plane surface at an angle of incidence of 40°, as shown. If the wave enters a region having a relative dielectric constant of 4 and then leaves for free space again, determine the direction of travel in the dielectric slab and in the free-space region into which it emits. (State all directions with respect to the surface normal.)

8-5. (a) If the index of refraction of glass is 1.5, what is the critical angle θ_c for an electromagnetic wave in glass?
  (b)

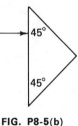

FIG. P8-5(b)

Explain what happens to the wave ray shown as it enters the glass prism (index of refraction = 1.5).

8-6. (a) Explain in your own words the surface wave as developed by Sommerfeld and others.

(b) A half-wave dipole is located 30 m above the ground. A receiving dipole 10 km distance is elevated to 5 m. Determine the field strength at the receiving antenna when the transmitting antenna carries a 2-A current at 100 MHz.

8-7. The index of refraction in the ionosphere is given by the equation

$$n = \sqrt{1 - \frac{Nq^2}{\omega^2 \epsilon_0 m}}$$

where $N \doteq$ electron density (electrons/m³)
$\quad q =$ charge of an electron (coulombs)
$\quad \omega =$ radial frequency ($2\pi f$)
$\quad m =$ mass of an electron
$\quad \epsilon_0 =$ dielectric constant of free space

(a) As one enters a medium of increasing ϵ_r, the beam:
  (i) Bends toward the line normal to the surface.
  (ii) Bends away from the line normal to the surface.
  (iii) Does not bend.

(b)

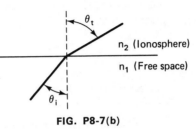

FIG. P8-7(b)

Circle the appropriate answer in the following table:

	n_2	θ_t
N increases	increases decreases	increases decreases
f increases	increases decreases	increases increases

8-8. From Fig. 8-19, what maximum frequency can be used at 8 P.M. if the skip zone is to be no greater than 500 mi?

8-9. (a) Describe the reason for obtaining a 6-dB gain in signal strength when the clearance over a plane earth equals one Fresnel zone, as shown in Fig. 8-27.
  (b) What is the loss (in dB) from free-space conditions when a knife edge having a clearance of -2 Fresnel zones is placed between two antennas? Obtain the same for a clearance of $+2$ Fresnel zones. (Assume the reflection coefficient of the knife edge to be zero.)

8-10. (a) On a smooth normal earth ($K = \frac{4}{3}$) it is desired to transmit between two points

30 mi apart at a frequency of 6 GHz. Obtain the radius of the first Fresnel zone at a point midway between the antennas.

(b) By using equation (8-12), draw a neat topographic profile of the $K = \frac{4}{3}$ earth over the 30-mi span.

(c) If a clearance of $\frac{1}{4}$ Fresnel radius is desired, what antenna heights are required? Assume equal antenna heights. Superimpose the microwave beam on your profile.

(d) Assuming smooth-sphere diffraction, what is the loss in dB from the free-space condition?

8-11. (a) Explain the strong interest in employing satellites for communication networks.

(b) Why are satellites so frequently located in the equatorial zone?

8-12. The refractive index of silica is $n_1 = 1.451$. An optical fiber such as that shown in Fig. 7-32 has a pure silica core and a doped silica cladding [$n_2 = n_1(1 - 0.01)$]. Because the rays can zigzag down the fiber at various angles, pulse dispersion occurs.

(a) For a step-index multimode fiber of length L, prove that the greatest differential time spread of a pulse is given by

$$\Delta T_{max} = \frac{L}{c} \frac{n_1}{n_2} (n_1 - n_2)$$

where n_1 = refractive index of the core

n_2 = refractive index of the cladding

c = velocity of light

(*Note:* This would be the time difference between the axially directed ray and the one zigzagging down the core at the critical angle.)

(b) Find the maximum transmitting pulse rate for a 1-km length of fiber with the given properties.

(c) What should the relationships be between n_1 and n_2 to obtain an optimum pulse rate?

8-13. Light is launched into an optical fiber having refractive indices for the cone and cladding of 1.5 and 1.49, respectively. Show that this results in an acceptance core half-angle 9.96°. What is the numerical aperture of this fiber and what does it represent?

nine

ANTENNA
FUNDAMENTALS

9-1 INTRODUCTION

In order to efficiently convert a guided wave to a free-space electromagnetic wave, a launching device called an *antenna* is attached to the guided wave structure. Often an antenna is located on a tower with the transmitting and receiving equipment at ground elevation for ready access to technical personnel.

An antenna serves two major functions. It acts as an impedance-matching device to match the impedance of the transmission line to that of free space, and it directs the radiation into the desired direction(s), or suppresses the radiation in other directions where it is not desired.

The same holds true for a receiving antenna as for a transmitting antenna. If, for instance, TV ghosting is a problem, where a secondary image is received due to the reception of a time-delayed reflected wave, a directional receiving antenna may be so positioned as to keep the undesired reflected wave from being detected. The direct signal may, however, still be received.

An antenna, whether receiving or transmitting, has identical electrical properties. The directional pattern and impedance properties of an antenna when receiving has the same directional pattern and imped-ance properties when employed as a transmitting antenna. This reciprocal property can be very significant, as it may result in a much more convenient method of impedance or pattern measurement. It may, for instance, be more convenient to measure the impedance of an antenna under a radiating condition, even though it may finally be used as a receiving antenna. At the same time, it might be desirable to measure the directional properties under receiving conditions. This is not to say, however, that transmitting antennas are built structurally the same as receiving antennas. Generally, transmitting antennas carry heavy currents and have high voltage gradients, which forces the designer to take certain precautions. Also, as in AM transmission, efficiency of the transmitting antenna is of much concern resulting in an electrically long antenna, whereas the receiving antenna is electrically quite short, but long enough to obtain a sufficient signal at the terminals for detec-tion in the receiver circuitry.

Even though the radiation pattern of an antenna, whether receiving or transmitting is identical, the current distribution on the antenna in the two cases may be different. To obtain some feeling for the fields from a transmitting antenna, for example polarization, it is helpful to know the approximate current distribution on an antenna.

It is possible to derive the expressions for the radiated fields from an antenna once the current distribution is known. We shall not follow this procedure, but the reader may refer to almost any basic antenna textbook for an introduction to this technique.

Consider an open-circuited transmission line which is opened up at points $\lambda/4$ from the open end to form a $\lambda/2$ dipole, as shown in Fig. 9-1(c). Since the cur-rent on an open-circuited line is zero at the end, and varies sinusoidally in magni-tude as one moves toward the generator, a similar current distribution can be expected to exist on the dipole as a first approximation.

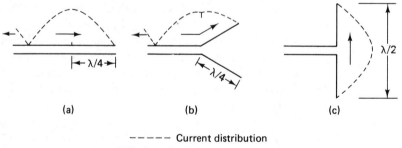

(a)   (b)   (c)

– – – – – Current distribution
FIG. 9-1 Evolution of a $\lambda/2$ dipole.

Since the antenna will radiate, the current distribution will differ somewhat from the sinusoidal form. The sinusoidal approximation is quite acceptable for pattern determination, but it may result in large errors in impedance information, particularly when the antenna terminals are located near the current nodal points. These current nulls tend to be filled out, owing to radiation, which results in a finite input impedance rather than the theoretical infinite impedance.

The current distribution can be obtained experimentally by using a small loop at the antenna surface, as indicated in Fig. 9-2. By moving the sampling shielded loop along the slot in the antenna, the surface current is monitored, since the tangential component of the magnetic field which cuts through the sampling loop is directly proportional to the surface-current density.

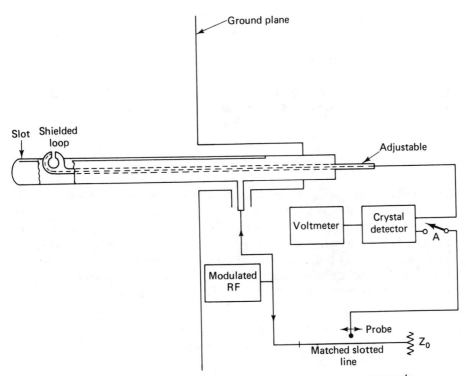

FIG. 9-2 Current phase and amplitude measurement system on a monopole antenna.

The phase measurement can be made with switch A closed. For a matched line, the phase varies uniformly along the line at a rate of βx or $(2\pi/\lambda)x$, where x is any point along the line. If the signal whose relative phase is to be measured is mixed with the reference signal from the slotted line probe by the crystal detector, the voltmeter will indicate the vector sum of the two signals. A reference is established at some convenient point on the antenna by locating the shielded loop at this point and by moving the slotted line probe until a minimum is indicated on the meter or where the signals are 180° out of phase. A variation in phase from this reference is obtained by measuring the displacement Δx of the slotted line probe to obtain a new null. The relative phase difference $\Delta\phi$ between any two points is given by

$$\Delta\phi = \frac{2\pi}{\lambda}\,\Delta x$$

where λ is the wavelength in the slotted line.

In this chapter we shall consider some of the fundamental properties of an antenna, such as:

1. Directional characteristics.
2. Gain and polarization.
3. Input impedance.

Although many of these properties can be rigorously derived, the mathematics is sufficiently difficult to prevent us from pursuing that approach.

9-2 SPHERICAL COORDINATE SYSTEM

When dealing with systems that have spherical symmetry, such as that encountered with fields radiating from an antenna, the spherical coordinate system is generally employed. It allows for the most simple form of mathematically representing the radiating fields. The spherical coordinate system is shown in Fig. 9-3, where the coordinates r, θ, and ϕ are mutually orthogonal (perpendicular to one another). When representing electromagnetic fields, these symbols are used as subscripts to denote the component or orientation of the particular field.

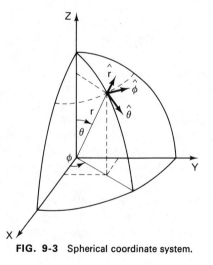

FIG. 9-3 Spherical coordinate system.

The unit vectors $\hat{r}$, $\hat{\theta}$, and $\hat{\phi}$ shown in Fig. 9-3 are often employed to describe the direction of a field. They have a magnitude of unity and therefore contribute solely to the directional quantity of field expression.

9-3 PHYSICAL PICTURE OF RADIATION

As the dipole is a common form of antenna and probably the one most thoroughly analyzed analytically and experimentally, we will chose it for illustrative purposes. It consists of two collinear wires, as shown in Fig. 9-1(c). As indicated in Fig. 9-1, a $\lambda/2$ dipole can be visualized as an open-circuited transmission line opened up a quarter-wavelength from the open end, with maximum current occurring at the antenna terminals and zero at the ends. The voltage standing wave is a maximum at the ends and a minimum at the terminals.

To visualize the energy flow near a dipole, consider a voltage pulse applied to the terminals. The electric field can be approximated by concentric circles, as shown in Fig. 9-4, with the magnetic field lines being concentric with the axis of the antenna.

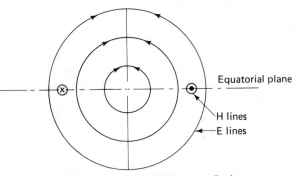

Equatorial plane

H lines
E lines

FIG. 9-4 Approximate fields near a dipole.

Initially, the fields tend to form a TEM mode, with no radial component. As the wave reaches the tip of the antenna, it observes an abrupt discontinuity and a large reflection occurs. Thus, little energy is radiated in the axial direction. Since no reflection occurs in the equatorial plane, energy does continue to flow out in the equatorial plane.

As the fields break away from the antenna, higher-order modes are formed in which the electric field forms closed loops. The field that results is shown in Fig. 9-5. It should be noted that a radial component is present, being largest near the polar axis. We will indicate a little later, in equation form, that the radial components of the fields attenuate at a more rapid rate as they propagate from the antenna than do the transverse components. Thus, at large distances from the antenna only transverse components are present.

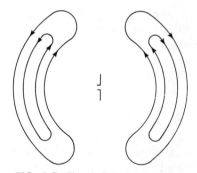

FIG. 9-5 Electric field near a dipole.

Another way of looking at radiation is to think of the current as continuously leaking off the wire into free space as one moves toward the antenna tip. This explains why the current on the antenna is reduced with distance along the wire. One can visualize this as the conduction current along the wire being replaced by an equal displacement current out of its surface.

9-4 ELEMENTAL OR SHORT ELECTRIC DIPOLE

Since any single conductor antenna may be considered as consisting of a large number of short conductors in series, the fields of such an antenna can be obtained by vectorally adding all the fields of the short conductors. For this reason, the fields of a short dipole having a uniform current distribution are often of interest, even if such a dipole does not exist. In actual practice, a short dipole has a sinusoidal current distribution. The significance of this will be considered at the end of this section. Let us consider an infinitesimally short length of wire (of zero thickness) which has a uniform current distribution I along its length as shown in Fig. 9-6. The length dl is very short compared to the wavelength.

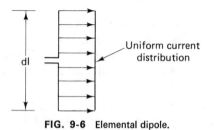

FIG. 9-6 Elemental dipole.

Since waves radiate from a source in a radial manner, it is convenient to place the dipole in a spherical coordinate system in order to describe the electric and magnetic fields. This is done in Fig. 9-7, where the dipole axis is placed in coinci-

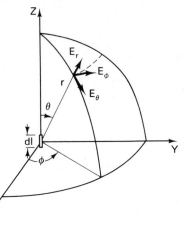

FIG. 9-7 Elemental dipole at the center of a spherical coordinate system.

dence with the z axis of the coordinate system. These fields are derived in almost any basic text on antennas, and are given by the expressions

$$E_\phi = 0 \tag{9-1a}$$

$$E_\theta = \frac{\eta I \, dl \, \sin \theta}{4\pi} \left(\frac{j\beta}{r} + \frac{1}{r^2} + \frac{1}{j\beta r^3} \right) e^{j(\omega t - \beta r)} \tag{9-1b}$$

$$E_r = \frac{\eta I \, dl \, \cos \theta}{4\pi} \left(\frac{2}{r^2} + \frac{2}{j\beta r^3} \right) e^{j(\omega t - \beta r)} \tag{9-1c}$$

$$H_\phi = \frac{I \, dl \, \sin \theta}{4\pi} \left(\frac{j\beta}{r} + \frac{1}{r^2} \right) e^{j(\omega t - \beta r)} \tag{9-1d}$$

$$H_\phi = 0 \tag{9-1e}$$

$$H_r = 0 \tag{9-1f}$$

Here E_θ is, for instance, the θ component of the electric field as referred to Fig. 9-7. The term $e^{j(\omega t - \beta r)}$ in the equations indicate that the waves travel in the radial r direction.

A study of the E_θ field reveals that the field has three components whose intensities decrease at different rates as r is increased. One term contains the factor $1/r$, one the factor $1/r^2$, and the third the factor $1/r^3$. This variation is shown in Fig. 9-8. When r becomes sufficiently large, the $1/r$ term is so much larger than the other two terms that they are negligible. When the $1/r$ term predominates, we are in the *far-field* region. When the other terms dominate, one is considered to be in the *near-field* region. Generally, the radiation patterns are considered only in the far-field region.

Thus, in the far field, where the terms in $1/r^2$ and $1/r^3$ can be neglected in favor of the terms in $1/r$, we are left with the two field components E_θ and H_ϕ. These are

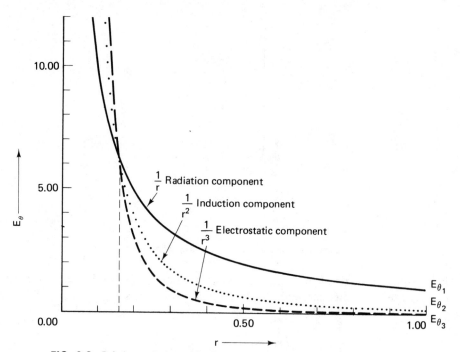

FIG. 9-8 Relative variation with distance of short-dipole electric field components.

given by

$$E_\theta = \frac{j\eta I \sin\theta}{2r} \frac{dl}{\lambda} \tag{9-2a}$$

$$H_\phi = \frac{jI \sin\theta}{2r} \frac{dl}{\lambda} \tag{9-2b}$$

The radiation terms of E_θ and H_ϕ are in time phase and are related by

$$\frac{E_\theta}{H_\phi} = \eta = 377\ \Omega \tag{9-3}$$

The *radiation pattern* or directional characteristics of an antenna is a diagram showing the relative intensity of the radiated field as a function of direction at a constant radius from the antenna. Since the three-dimensional plot is difficult to draw, the patterns are usually drawn for two planes: one called the *vertical pattern*, which lies in the plane containing the antenna axis as shown in Fig. 9-9(a), and one called the *horizontal pattern*, which lies in the plane normal to the antenna axis.

The maximum field strength occurs at $\theta = 90°$ or at a position perpendicular to the antenna axis. At this particular point, the field is in the vertical direction and the wave is said to be *vertically polarized*. In general, the electric field is θ polarized.

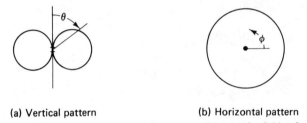

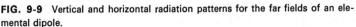

(a) Vertical pattern   (b) Horizontal pattern

FIG. 9-9 Vertical and horizontal radiation patterns for the far fields of an elemental dipole.

The magnetic field from equation (9-2b) is seen to be perpendicular to the electric field (in the ϕ direction) and forms the same radiation pattern as that of the E field. Since the units of the electric field are V/m and the units of the magnetic field are A/m, when multiplying $\mathbf{E} \times \mathbf{H}$, the units turn out to be W/m², or power density. Since the electric and magnetic fields need not be always in phase, the real power density at any point is given by

$$\mathcal{P} = \mathbf{E} \times \mathbf{H}^* \qquad \text{W/m}^2 \qquad (9\text{-}4)$$

where the $\times$ is known as the cross product. Rotating $\mathbf{E}$ toward $\mathbf{H}$ and noting the direction that a normal screw would move gives the direction of power flow. The $*$ stands for the complex conjugate of the $\mathbf{H}$.

For the case of the elemental dipole, substituting equations (9-2) into equation (9-4) and noting that $H_\phi = E_\theta/\eta$:

$$\mathcal{P} = E_\theta \times H_\phi^*$$

$$\mathcal{P} = E_\theta \frac{E_\theta^*}{\eta^*} \hat{r} = \frac{E_\theta^2}{\eta} \hat{r} \qquad (9\text{-}5)$$

where the $\hat{r}$ denotes the direction of power flow in the radial direction. $\eta^* = \eta$ since η is real (377 Ω).

EXAMPLE 9-1

Let us find the magnitude of the field intensity and power density of a signal 1 mi distance from an elemental dipole $\lambda/360$ long carrying a 1-A current. Assume a position perpendicular to the antenna.

Solution:

$$|E_\theta| = \frac{\eta I \sin \theta}{2r} \frac{dl}{\lambda}$$

$$= \frac{377 \times 1 \times 1 \times \lambda/360}{2 \times 5280 \text{ ft} \times 0.3048 \text{ m/ft} \times \lambda}$$

$$= 0.325 \text{ mV/m}$$

$$\mathcal{P} = \frac{E_\theta^2}{\eta}\hat{r}$$

$$= \frac{(0.325 \times 10^{-3})^2}{377}\hat{r}$$

$$= 0.28 \times 10^{-9} \text{ W/m}^2$$

in the direction normal to the antenna axis.

Since an antenna radiates energy, and therefore dissipates energy from a source, it is customary to associate this loss of power with a *radiation resistance*. If the total power radiated away from the antenna is P_t and the antenna current is I, the radiation resistance is

$$R_{\text{rad}} = \frac{P_t}{I^2}$$

The radiation resistance of an elemental dipole is given by

$$R_{\text{rad}} \text{ (elem. dipole)} = 80\pi^2 \left(\frac{dl}{\lambda}\right)^2 \tag{9-6}$$

For the example just discussed, where $dl = \lambda/360$, the radiation resistance is

$$80\pi^2 \left(\frac{\lambda/360}{\lambda}\right)^2 = 0.00609 \ \Omega$$

If this antenna is required to radiate 1 kW of power, the corresponding driving current must be

$$I = \sqrt{\frac{P_t}{R_{\text{rad}}}} = \sqrt{\frac{1000}{0.00609}} = 405 \text{ A}$$

This result indicates a major disadvantage of elemental dipoles, in that large currents are required for appreciable radiated power. For this reason longer dipoles are employed.

An actual short dipole can be conceived as made up of a short length of open-circuited transmission line with the wire bent to 90°, as shown in Fig. 9-10. In this case the antenna current is no longer uniform along the antenna length but can be considered to be linearly decreasing toward the antenna ends.

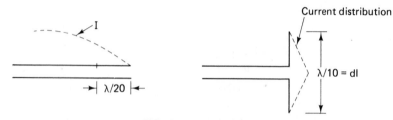

FIG. 9-10 Short dipole.

The antenna radiation resistance is effectively one-quarter that of the elemental dipole and is therefore equal to

$$R_{\text{rad}} = 20\pi^2 \left(\frac{dl}{\lambda}\right)^2 \tag{9-7}$$

The radiation patterns are identical to those shown in Fig. 9-9 for the elemental dipole. The beamwidth of a radiation pattern is usually taken to be the angular width of the main beam between the half-power points, or $1/\sqrt{2}$ electric or magnetic field intensity points. Thus, the beam width is 90° for a short dipole antenna, since, referring to equation (9-2), $\sin 45° = 1/\sqrt{2}$.

The input impedance to a short dipole consists of a small radiation resistance in series with a small capacitance (large capacitive reactance). This is consistent with the analogy to the open-circuited transmission line.

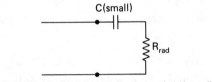

FIG. 9-11   Equivalent circuit of a center-fed short dipole.

9-5   HALF-WAVELENGTH DIPOLE

To improve the radiation efficiency of an antenna, a longer antenna is used. The half-wavelength dipole is a typical example. By considering the center-fed antenna as an open-circuited transmission line that has been opened up, the current distribution will appear as shown in Fig. 9-12. With a sinusoidal current distribution,

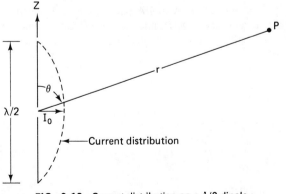

FIG. 9-12   Current distribution on a $\lambda/2$ dipole.

the far fields of a half-wavelength dipole at a distance r from the antenna are

$$E_\theta = \frac{jI_0\eta}{2\pi r} \; \frac{\cos(\pi/2 \cos\theta)}{\sin\theta} e^{-j\beta r} \qquad (9\text{-}8a)$$

$$H_\phi = \frac{E_\phi}{\eta} \qquad (9\text{-}8b)$$

The far-field patterns appear as shown in Fig. 9-13.

(a) Vertical pattern   (b) Horizontal pattern

FIG. 9-13   Vertical and horizontal radiation patterns of a vertical half-wavelength dipole.

The radiation pattern of a half-wavelength dipole is only slightly more directive than the elemental dipole, with a 3 dB beam width of 78° rather than 90°. The real advantage of the longer dipole is that its radiation resistance is much larger than that of the electrically short antenna. When the antenna is exactly a half-wavelength long, it also has a slight capacitive component in addition to the 73 Ω resistance. This reactance is usually eliminated by slightly shortening the antenna by about 5%. This results in a series resonant situation with an input resistance of 67 Ω.

The diameter of the antenna conductor has a great bearing also on the input impedance. More will be said on this matter in Section 9-12.

EXAMPLE 9-2

A $\lambda/2$ dipole in free space is driven with 1 A (rms) at the antenna terminals (this is the magnitude I_0 of the current loop). Determine the angle θ for maximum field strength and compute the field strength and power density at a point 1 mi from the antenna at the corresponding angle.

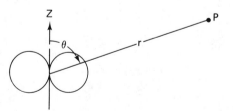

FIG. 9-14   Radiation pattern for antenna in Example 9-2.

Solution: The maximum field strength occurs at an angle of $\theta = 90°$ from the antenna axis. The electric field intensity magnitude can be computed by employing equation (9-8a), where

$$|E_\theta| = \frac{I_0\eta}{2\pi r} \frac{\cos{(\pi/2 \cos\theta)}}{\sin\theta}$$

At $\theta = 90°$, $I_0 = 1$ A, $r = 5280 \times 0.3048$ m.

$$|E_\theta| = \frac{1 \times 377}{2\pi \times 5280 \times 0.3048} \frac{\cos 0}{1}$$

$$= 37.3 \text{ mV/m}$$

The field is vertically polarized at this point and the power density is, from equation (9-5),

$$E_\theta \times H_\phi^* = \frac{E_\theta^2}{\eta} \hat{r}$$

$$= \frac{(37.3 \times 10^{-3})^2}{377} = 3.7 \ \mu\text{W/m}^2$$

in the direction normal to the antenna axis.

9-6   LONG LINEAR ANTENNAS

Increasing the length of an antenna causes the radiation resistance to increase, but also results in multilobing when a length of 1 wavelength is exceeded. The radiation patterns of some center-fed long vertical dipoles are shown in Fig. 9-15.

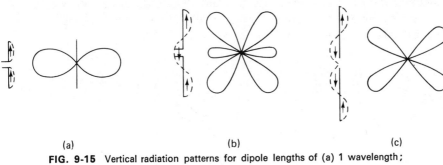

(a)   (b)   (c)

FIG. 9-15 Vertical radiation patterns for dipole lengths of (a) 1 wavelength; (b) 1½ wavelengths; (c) 2 wavelengths. The current distribution assumed is shown in each case.

When the feed point is changed, the radiation patterns also change. This is due to the asymmetry caused in the standing-wave currents set up on the antenna. Figure 9-16 shows the radiation pattern of linear antennas having various locations for feed points.

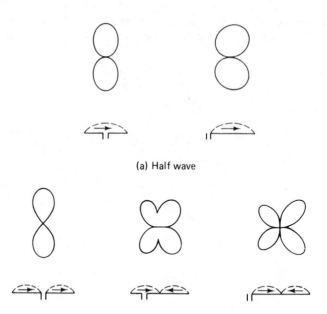

(a) Half wave

(b) Full wave

FIG. 9-16   Radiation patterns of linear antennas for various feed-point locations.

Nonresonant Antennas

Thus far we have considered resonant antennas with sinusoidal current distributions. Rather than allowing such a standing wave to exist on an antenna, one end can be terminated to permit only an incident wave on the antenna. One such antenna is the nonresonant antenna shown in Fig. 9-17(a).

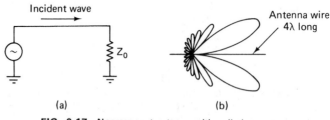

FIG. 9-17   Nonresonant antenna with radiation patterns.

Ignoring the effect of any ground, the far field tends to have its lobes directed in the direction of the traveling wave on the antenna. Figure 9-17(b) shows the far field of a nonresonant antenna 4 wavelengths long having negligible attenuation. Losses in such an antenna modify the radiation pattern by filling in the minima between the lobes. Each half-wavelength contributes another lobe to the radiation pattern. One rather broad-band antenna employing nonresonant lines is the rhombic antenna shown in Fig. 9-18. The antenna wires are held up above ground by supporting poles.

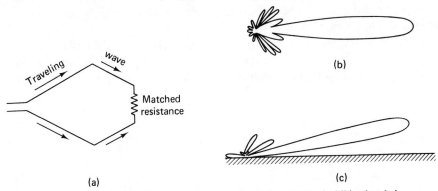

FIG. 9-18 (a) Rhombic antenna above a ground place with typical (b) azimuthal, and (c) vertical radiation patterns.

9-7 ANTENNAS ABOVE A GROUND PLANE

When an antenna is located near a ground plane, the fields may take on noticeably different patterns and the input impedance can be drastically altered. Assuming the ground plane to be a good conductor, the tangential component of the electric field must be zero at its surface. If we consider a horizontal antenna above a ground plane as that shown in Fig. 9-19, this boundary condition is fulfilled by causing the reflected wave to undergo a phase shift of 180° at the point of reflection.

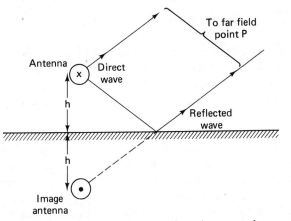

FIG. 9-19 Method of images applied to horizontal antenna above a ground plane.

The same field would be obtained at point P if the ground plane were replaced by an image antenna that has its current distribution reversed by 180° and placed a distance h below where the ground plane was previously located. The boundary condition of zero electric tangential field would be met, and the problem is greatly simplified. A little later, when discussing antenna arrays, the fields of such an antenna will be given.

Similarly, the method of images can be used for a vertical antenna above a ground plane. In this case the ground plane is replaced by an image antenna of equal length with a current distribution as indicated in Fig. 9-20. Referring to Fig.

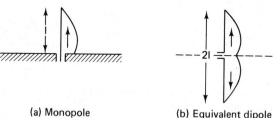

(a) Monopole   (b) Equivalent dipole

FIG. 9-20   Method of images applied to a monopole antenna.

9-20, we can see that a $\lambda/4$ monopole will result in the same radiation pattern as that given for a $\lambda/2$ dipole antenna in free space, except that it will be only half the free-space pattern, as the earth "cuts off" the other half. The pattern of the $\lambda/4$ monopole antenna is shown in Fig. 9-21.

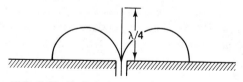

FIG. 9-21   Radiation pattern of a $\lambda/4$ monopole.

For a given base current, the total radiated power for a monopole is only one-half that for the antenna-plus-image antenna in free space, and therefore the radiation resistance is only half as great as that calculated on the antenna-plus-image antenna basis, or 36.5 Ω. There is also a reactance of $j21$ Ω for a $\lambda/4$ monopole.

Monopoles of lengths greater than one-quarter wavelength behave similarly to that of dipoles of lengths greater than one-half wavelength. Figure 9-22 shows the radiation patterns of monopoles of heights greater than one-quarter wavelength. It should be noted that multilobing starts when the monopole height begins to exceed approximately one-half wavelength.

When an irregular ground plane is used, the resulting radiation patterns can be radically altered. Figure 9-23 shows some patterns of a monopole located at various positions on an automobile.

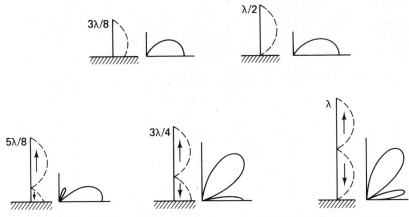

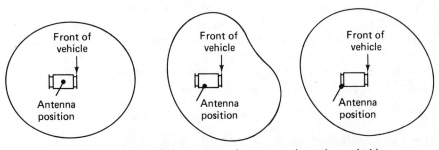

FIG. 9-22 Vertical radiation patterns for monopoles of various lengths over a perfect earth.

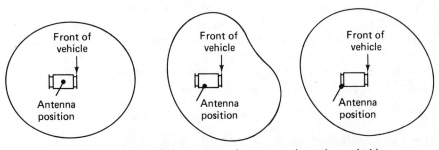

FIG. 9-23 Horizontal radiation patterns of an antenna located at typical locations on an automobile.

9-8 SMALL LOOP ANTENNA

A loop antenna consists of a circular section of wire which is considered small if the radius of the loop is very small compared to a wavelength. Let us consider the fields of a loop that is oriented in a spherical coordinate system as depicted in Fig. 9-24.

If the radius $a \ll \lambda$, then the far fields at a radius r are given by

$$E_\phi = \frac{\eta \pi I \sin \theta}{r} \frac{A}{\lambda^2} \qquad (9\text{-}9a)$$

$$H_\theta = \frac{\pi I \sin \theta}{r} \frac{A}{\lambda^2} \qquad (9\text{-}9b)$$

where A is the area of the loop or πa^2. Just as in the case of the short dipole, both fields are independent of ϕ and the far-field pattern has a doughnut form. The field pattern has the same shape as that of the short dipole which has its axis coincident

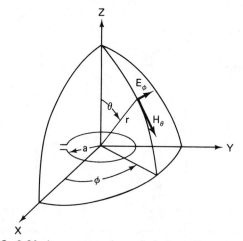

FIG. 9-24   Loop antenna in a spherical coordinate system.

with the loop axis, with the vector directions of the electric and magnetic components interchanged (Fig. 9-25). These antennas see application in direction finding and are often employed to monitor the habits of animals, placing them as transmitting antennas around the animals' necks.

With the loop used as a receiving antenna, the received voltage obtainable by suitably placing the antenna in an electric field E_ϕ is given by

$$V = \frac{2\pi}{\lambda} NAE_\phi \sin \theta \tag{9-10}$$

where N represents the number of turns in the loop.

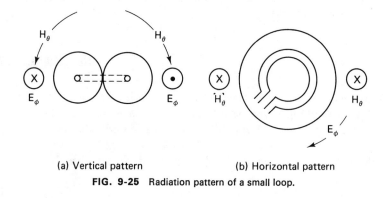

(a) Vertical pattern   (b) Horizontal pattern

FIG. 9-25   Radiation pattern of a small loop.

9-9   ANTENNA GAIN

Antennas in general have two functions to perform. One is to impedance-match the transmission line to that of free space, and the other is to direct the energy into desired directions. The latter has already been dealt with in a descriptive manner and is more exactly specified in terms of power gain, directive gain, or directivity.

The power radiated per unit area, commonly called the *power density*, $\mathcal{P}$, is given by equation (9-5). For a field having an electric power density of E volts per meter, the corresponding power density is given by

$$\mathcal{P} = \frac{E^2}{\eta} \quad \text{W/m}^2 \tag{9-11}$$

Let us consider an antenna radiating a total transmitted power of P_t resulting in a power density of $\mathcal{P}$ at some point P, as shown in Fig. 9-26. Assume the distance to the observation point P from the antenna to be r. The transmitted power is very difficult to measure and it differs from the power delivered to the antenna by the heat losses in the antenna element. For 100% efficient antennas, however, the two are equivalent, and it allows one to develop on a theoretical basis the gain of an antenna.

FIG. 9-26   Point P a distance r from an antenna.

The *directive gain*, *gd*, in a given direction is defined as the ratio of the power density in that direction at some distance r to the power density that would be radiated at the same distance r by an isotropic antenna radiating the same total power. This can be expressed as

$$g_d = \frac{\mathcal{P}}{P_t/4\pi r^2} \tag{9-12}$$

where $P_t/4\pi r^2$ is the power density at a distance r from an isotropic antenna radiating a total power of P_t.
Equation (9-12) can be rewritten as

$$g_d = 4\pi r^2 \frac{\mathcal{P}}{P_t} \tag{9-13}$$

When expressed in decibels, the directive gain is denoted by G_d, where

$$G_d = 10 \log g_d \tag{9-14}$$

The *directivity*, *D*, of an antenna is the maximum value of the directive gain of an antenna. The directivity or maximum directive gain of an elementary dipole is 1.5, or equivalent to $10 \log 1.5 = 1.76$ dB. For the half-wave dipole, the directivity is 1.64, or 2.15 dB.

The *power gain* takes into account the ohmic losses of the antenna. It is very similar to the directive gain in that it is the ratio of the power density at some point P a distance r from the antenna to the power density radiated at a distance r from an isotropic antenna radiating the same total power as that delivered to the antenna under consideration. The power gain, g_p, is thus defined by

$$g_p = \frac{\mathcal{P}}{P_d/4\pi r^2} \qquad (9\text{-}15)$$

where P_d is the power delivered to the antenna.

Dividing equation (9-15) by equation (9-12), we see that the ratio of power gain to directive gain is a measure of the efficiency of the antenna:

$$\frac{g_p}{g_d} = \frac{P_t}{P_d} \qquad (9\text{-}16)$$

9-10 EFFECTIVE AREA OF AN ANTENNA

The *effective area* or *receiving cross section of an antenna*, a term frequently employed when considering receiving antennas, is defined as the ratio of power available at the receiving terminals of the antenna to the power density of the appropriately polarized incident wave:

$$A = \frac{P_r}{\mathcal{P}} \qquad (9\text{-}17)$$

where P_r = received power

$\mathcal{P}$ = power density of the incident wave at the receiving antenna location

A = effective area of the antenna

Expressing equation (9-17) somewhat differently, the received power is equal to the power density at the antenna times the effective area of the antenna. If, for example, the power density at the receiving antenna location is 1 mW/m² and the effective area of the receiving antenna is 1/2 m², the total received power at the terminals is

$$\frac{1 \text{ mW}}{\text{m}^2} \times \frac{1}{2} \text{ m}^2 = \frac{1}{2} \text{ mW}$$

It can be shown that the effective area A of an antenna is given by

$$A = \frac{\lambda^2}{4\pi} \times g_d \qquad (9\text{-}18)$$

Equation (9-18) assumes that no losses occur in the receiving antenna and that all the available power is delivered to the load. This means that the antenna is properly matched and has the proper polarization.

EXAMPLE 9-3

Calculate the effective area of a $\lambda/2$ dipole at 50 MHz.

Solution:

$$v = f\lambda$$

$$\lambda = \frac{3 \times 10^8}{50 \times 10^6} = 6 \text{ m}$$

$$A = \frac{\lambda^2}{4\pi} \times g_d = \frac{6 \times 6}{4\pi} \times 1.64 = 4.7 \text{ m}^2$$

9-11 FREE-SPACE PATH LOSS

We can now develop the expression for the power that will be received by a receiving antenna having a gain g_{dr} when the total power transmitted from a transmitting antenna having a gain g_{dt} is P_t. Consider two antennas in free space separated by a distance r, as shown in Fig. 9-27. The power density $\mathcal{P}$ at a distance r from the

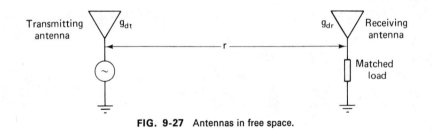

FIG. 9-27 Antennas in free space.

transmitting antenna is given by

$$\mathcal{P} = \frac{g_{dt}P_t}{4\pi r^2} \quad \text{W/m}^2 \tag{9-19}$$

Expression (9-19) is another form of equation (9-13).

The receiving antenna, having an effective area A_r, will capture a total power of

$$P_r = A_r\mathcal{P} \tag{9-20}$$

Since from equation (9-18),

$$A_r = \frac{\lambda^2}{4\pi} g_{dr}$$

expression (9-20) can be rewritten as

$$P_r = \mathcal{P} \frac{\lambda^2}{4\pi} g_{dr} \tag{9-21}$$

Substituting equation (9-19) into (9-21), we obtain for the received power

$$P_r = \frac{P_t g_{dt} g_{dr} \lambda^2}{(4\pi r)^2} \qquad (9\text{-}22)$$

The free-space loss defined as the ratio of the transmitted power to the power obtainable at the receiving antenna terminals is given by

$$\frac{P_t}{P_r} = \left(\frac{4\pi r}{\lambda}\right)^2 \Big/ g_{dt} g_{dr} \qquad (9\text{-}23)$$

The free-space loss in dB is thus given by

$$\text{FSL (free-space loss)} = 10 \log \left(\frac{4\pi r}{\lambda}\right)^2 \Big/ g_{dt} g_{dr} \qquad \text{dB} \qquad (9\text{-}24)$$

For r in miles and frequency in MHz, equation (9-24) can be simplified to

$$\text{FSL} = 36.6 + 20 \log r + 20 \log f - G_{dt} - G_{dr} \qquad \text{dB} \qquad (9\text{-}25)$$

G_{dt} and G_{dr} are the gains in dB of the transmitting and receiving antennas, respectively. If isotropic antennas are employed, equation (9-25) reduces to

$$\text{FSL (isotropic antennas)} = 36.6 + 20 \log r + 20 \log f \qquad \text{dB} \qquad (9\text{-}26)$$

(where r is in miles and f is in MHz). This expression is the free-space loss referred to in Fig. 8-27.

For r in kilometers and frequency in GHz, equation (9-25) converts to

$$\text{FSL} = 92.4 + 20 \log r + 20 \log f - G_{dt} - G_{dr} \qquad (9\text{-}27)$$

(where r is in km and f is in GHz).

A frequently used tool in describing antenna performance is the *effective isotropically related power* (EIRP), or the power in dBm or dBW over an isotropic antenna that is radiated. Referencing the power of equation 9-23 to 1 W (dBW), we obtain

$$10 \log P_r = 10 \log P_t + G_t + G_r + 20 \log \frac{\lambda}{4\pi r} \qquad (9\text{-}28)$$

The combination of the first two terms on the right-hand side of equation 9-28 is known as the EIRP:

$$\text{EIRP} = P_t \text{ (in dBW)} + G_t \qquad (9\text{-}29)$$

EXAMPLE 9-4

A transponder in a satellite has an output of 5 W; with negligible line losses and an antenna gain of 25 dB, what is the EIRP of the main beam?

Solution:

$$5 \text{ W} = 10 \log \frac{5}{1} = 7 \text{ dBW (or 37 dBm)}$$

$$\text{EIRP} = 7 \text{ dBW} + 25 \text{ dB} = 32 \text{ dBW (or 62 dBm)}$$

Any line losses would decrease the EIRP by this loss.

Figure 9-28 shows the free-space loss for the three frequencies 4 GHz, 6 GHz, and 11 GHz. Although the FSL increases with frequency for isotropic antennas, actual antennas may increase in gain for a fixed physical size to more than compensate for this loss.

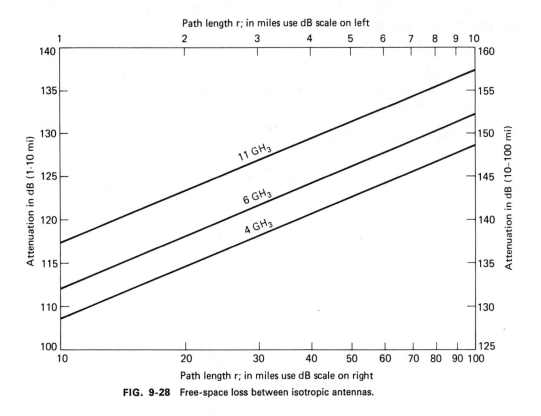

FIG. 9-28 Free-space loss between isotropic antennas.

EXAMPLE 9-5

Consider two 100% efficient $\lambda/2$ dipoles spaced 20 km apart and operating at 50 MHz. If 10 W is delivered to the transmitting antenna, determine the power that will be obtained at the receiving terminals.

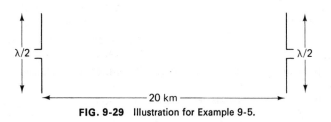

FIG. 9-29 Illustration for Example 9-5.

Solution: The gain of a $\lambda/2$ dipole is 1.64, and the wavelength at 50 MHz is 6 m. Substituting into equation (9-22),

$$P_r = \frac{10 \times 1.64 \times 1.64 \times 6^2}{(4\pi \times 20 \times 10^3)^2} = 15.3 \text{ nW}$$

9-12 INPUT IMPEDANCE OF CYLINDRICAL ANTENNAS

Assuming the current distribution on an antenna to be similar to that on an open-circuited transmission line is a reasonably valid approximation except in regions near the current minimums. The current distribution around these regions depends very much upon the diameter of the antenna. Figure 9-30 shows a typical current distribution on an antenna 1 wavelength long as well as the sinusoidally distributed current present on an open-circuited line.

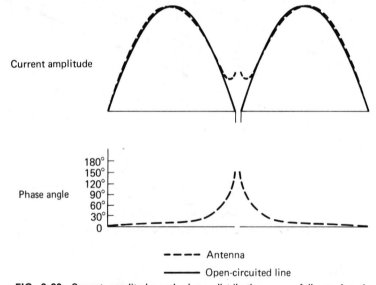

FIG. 9-30 Current amplitude and phase distribution on a full-wavelength antenna of finite diameter compared to the current distribution on an open-circuited line.

Since the antenna input impedance is the ratio of the voltage to current at the antenna terminals, the input impedance becomes very much a function of the diameter of the antenna. Several researchers have developed expressions for the input impedance of cylindrical antennas, each following a different approach. Notable among these are the results derived by Schelkunoff, Hallen, and King.[1] Rather than presenting these theoretical results, which very closely approximate

[1] Refer, for instance, to I. D. Kraus, *Antennas* (New York: McGraw-Hill Book Company, 1950), Chap. 9.

the real situation, we shall consider some measured data. The theoretically determined impedances differ from measured quantities because the capacitance between the monopole terminal and ground and the antenna end effects are not taken into account in the theoretical development. Actual measurements are usually performed on a monopole in order to reduce interference effects from the measurement equipment.

As illustrated in Fig. 9-31, each antenna will have a slightly different input impedance, with Fig. 9-31(b) being closer to the theoretical model. Figure 9-32 shows typical input impedance curves of two monopole antennas having length/diameter ratios ($l/2a$) of 10 and 236. Comparing the two curves, it is apparent that the thicker antenna has a decreased impedance variation with frequency. This phenomenon makes the thick antenna useful in broad-band applications.

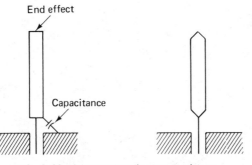

FIG. 9-31 Some monopole antenna shapes.

An antenna is considered to be resonant when the input impedance is purely resistive. At low frequencies, where the antenna is electrically short, the impedance is chiefly capacitively reactive with little series resistance. The first series resonance occurs when the length of the monopole is slightly less than 1/4 wavelength. As the frequency is increased, the antenna length electrically increases and the second resonance (parallel resonance) occurs at slightly less than 1/2 wavelength.

It is common practice to operate antennas near or at the resonance points. For this reason the driven element on a Yagi antenna is usually slightly shorter than one-half wavelength. As the length/radius ratio goes to infinity, the resonant points approach an integral number of one-quarter wavelengths.

To obtain the input impedance of a dipole in free space, the results in Fig. 9-32 are mutiplied by a factor of 2. The diameter of the antenna has little effect on the radiation patterns.

Top-Hat Antenna

At low frequencies where the wavelengths become very long, it becomes very expensive to build vertical antennas of a reasonable electrical length. To reduce this problem, "top loading" is often employed, whereby the effective electrical length

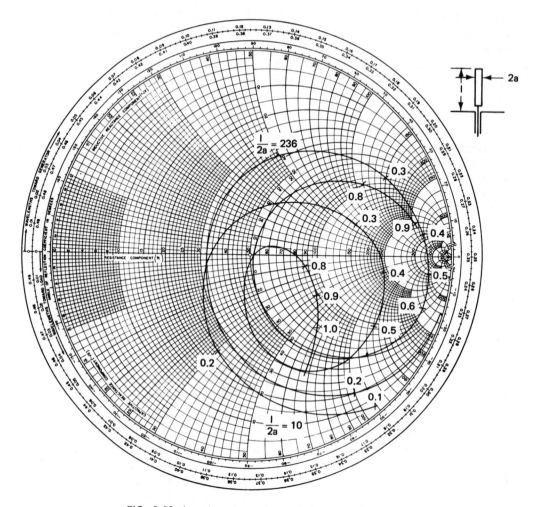

FIG. 9-32 Input impedance of a monipole antenna for length/diameter ratios of 10 and 236.

is increased by running wires horizontally from the upper end of the monopole. The horizontal section also radiates but proportionally less than the vertical section, since the current distribution is sinusoidal as shown in Fig. 9-33. The input impedance will also be less capacitive than would be the case if the top hat were not placed on the vertical section. Figure 9-34 shows a typical top-load antenna installation.

To reduce power losses due to ohmic resistance of the ground, radial copper wires are often extended radially from the base of the antenna. In a standard AM broadcast station, as many as 120 radial wires extending to $\lambda/4$ from the tower base are used. They are generally buried for protection, 1 or 2 ft below the surface. Ground resistance can be decreased in this way from 50 Ω to 1 Ω. Since the electrical connection between the radial system and the earth is largely capacitive, the wires may also lay on the surface or even be supported on poles above the surface.

Current distribution

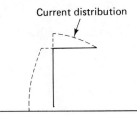

FIG. 9-33 Top-loaded antenna.

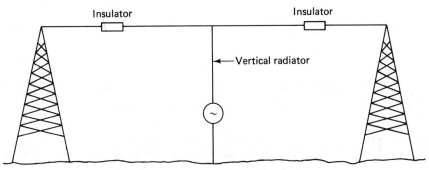

FIG. 9-34 Top-loaded antenna installation.

9-13 SELF- AND MUTUAL IMPEDANCES

When an antenna is isolated, that is, remote from any other antenna, ground, or object, the input impedance is equal to the self-impedance of the antenna. When locating another antenna or structure near an antenna, the input impedance is no longer solely the self-impedance but is also affected by the coupled fields from the other antennas or structures which have currents flowing in them.

One can write a set of simultaneous equations describing this situation. Let us consider two antennas having terminal voltages and currents V_1, V_2 and I_1, I_2, respectively. For two antennas such as that shown in Fig. 9-35, the input voltages

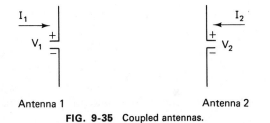

FIG. 9-35 Coupled antennas.

and currents can be related by the expression

$$V_1 = Z_{11}I_1 + Z_{12}I_2 \qquad (9\text{-}30a)$$

$$V_2 = Z_{21}I_1 + Z_{22}I_2 \qquad (9\text{-}30b)$$

where Z_{11} is the impedance of antenna 1 with antenna 2 present and open-circuited:

$$Z_{11} = \frac{V_1}{I_1}\bigg|_{I_2=0} \qquad (9\text{-}31)$$

Similarly, Z_{22} is the impedance of antenna 2 with antenna 1 present and open-circuited.

$$Z_{22} = \frac{V_2}{I_2}\bigg|_{I_1=0} \qquad (9\text{-}32)$$

For media not having variable permeability and dielectric constant, the mutual impedances

$$Z_{12} = Z_{21}$$

The mutual impedance between two antennas on a ground plane (see Fig. 9-36) can be experimentally determined by the following procedure:

1. Set $I_2 = 0$ by placing a short $3\lambda/4$ from antenna 2 terminals.
2. Measure the input impedance to antenna 1 with antenna 2 open-circuited.

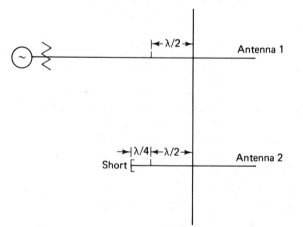

FIG. 9-36   Mutual impedance measurement system.

This is equivalent to Z_{11} as

$$Z_{11} = \frac{V_1}{I_1}\bigg|_{I_2=0}$$

The impedance measurement can follow one of the techniques outlined in Chapter 5.

3. Since

$$Z_{21} = Z_{12} = \frac{V_2}{I_1}\bigg|_{I_2=0} \qquad (9\text{-}33)$$

from equation (9-30b), we can substitute for I_1 from step 2:

$$I_1 = \frac{V_1}{Z_{11}}\bigg|_{I_2=0}$$

Thus,

$$Z_{21} = Z_{12} = Z_{11}\frac{V_2}{V_1}\bigg|_{I_2=0} \qquad (9\text{-}34)$$

The voltage ratio V_2/V_1 is measured with a vector voltmeter with antenna 2 open-circuited as before.

The mutual impedance is determined by multiplying the measured values

$$Z_{11} \times \frac{V_2}{V_1}\bigg|_{I_2=0}$$

Figure 9-37 shows the mutual impedance between two thin parallel half-wave antennas remote from earth. The impedance depends upon the antenna spacing with the magnitude being greater for small spacings.

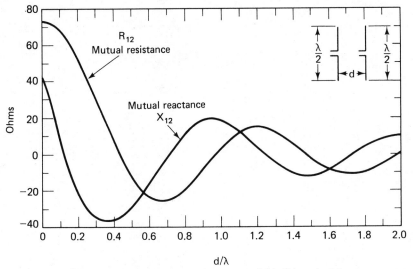

FIG. 9-37 Mutual impedance curves for two parallel half-wavelength antennas.

For an application of mutual impedance, let us obtain the input resistance of a horizontal half-wavelength-long dipole a distance h above a ground plane, as depicted in Fig. 9-38. Employing the method of images (see Fig. 9-19), the ground plane in such an arrangement can be replaced by a dipole a distance h below the ground-plane surface. This image antenna carries an equal but 180° out-of-phase

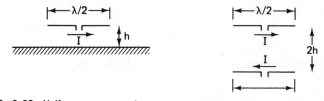

FIG. 9-38 Half-wave antenna above a ground plane and the image equivalent.

current with respect to the driven antenna. In such a situation, where I_2 (the image antenna current) $= -I_1$ (the driven antenna current), equation (9-30a) reduces to

$$V_1 = Z_{11}I_1 - Z_{12}I_1$$

or

$$\frac{V_1}{I_1} = Z_{11} - Z_{12} \tag{9-35}$$

The input resistance is thus given by

$$R_{in} = R_{11} - R_{12} \tag{9-36}$$

where R specifies the real part of the impedance only. Since the self-resistance of a $\lambda/2$ dipole is 73 Ω,

$$R_{in} = 73 - R_{12}$$

Obtaining R_{12} from Fig. 9-37, we obtain the input resistance of the driven element as illustrated in Fig. 9-39.

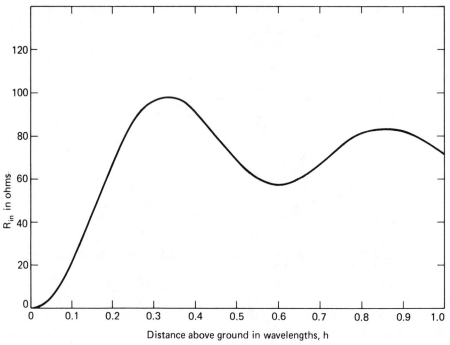

Distance above ground in wavelengths, h

FIG. 9-39 Input resistance of a horizontal half-wavelength antenna at a height h wavelengths above a ground plane.

Another example of a coupled antenna is the folded dipole shown in Fig. 9-40. It can be shown that the terminal impedance is four times the input impedance of that of a single element. Thus, the $\lambda/2$ folded dipole has an input impedance of approximately $4 \times 70 = 280\,\Omega$. Such a folded dipole is a reasonably good match to a $300\,\Omega$ twin lead commonly employed in TV reception. In addition, this arrangement increases the bandwidth of the antenna.

FIG. 9-40 Folded dipole.

9-14 ANTENNA ARRAYS

When it is desired to increase the directivity over that obtained by a single antenna, antenna arrays are used. An antenna array consists of a number of similar antennas oriented in a particular fashion whose directivity depends upon the phase interference of the electromagnetic waves sent from the various individual antennas. Consider the two-element array shown in Fig. 9-41, in which the antennas are isotropic radiators with currents as indicated.

At a point P that is sufficiently remote from the array, the phase difference due to the difference in propagation distances is

$$\beta d \cos \phi$$

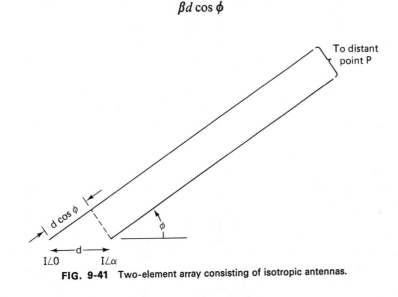

FIG. 9-41 Two-element array consisting of isotropic antennas.

where β = phase constant in free space
d = distance separating the antennas
ϕ = angle between the array axis and the line joining point P and the array

The total phase difference between the radiations from the two antennas at point P will be

$$\psi = \beta d \cos \phi + \alpha \tag{9-37}$$

where α is the phase angle by which the current in the second antenna leads the current in the first antenna.

As far as the magnitude of the individual fields from the antenna is concerned, the inverse distance terms $(1/r)$ in the far-field expressions can be considered to be identical. If the field strengths due to one antenna alone is E_0, the magnitude of the field strength of the two antennas will be given by

$$
\begin{aligned}
|E_T| &= |E_0 + E_0 e^{j\psi}| \\
&= |E_0| |1 + e^{j\psi}| \\
&= |E_0| \sqrt{(1 + \cos \psi)^2 + \sin^2 \psi} \\
&= |E_0| \sqrt{1 + 2 \cos \psi + \cos^2 \psi + \sin^2 \psi} \tag{9-38}
\end{aligned}
$$

but

$$\cos^2 \psi + \sin^2 \psi = 1$$

and

$$\cos^2 \psi = \frac{1 + \cos \psi}{2}$$

so

$$
\begin{aligned}
|E_T| &= |E_0| \sqrt{2(1 + \cos \psi)} \\
&= |E_0| \sqrt{2^2 \cos^2 \psi/2} \\
&= 2|E_0| \cos \frac{\psi}{2} \\
&= 2|E_0| \cos \left(\frac{\pi d \cos \phi}{\lambda} + \frac{\alpha}{2} \right) \tag{9-39}
\end{aligned}
$$

The radiation patterns for two omnidirectional antennas with equal excitations, but of various spacings and phasings, are shown in Fig. 9-42. Figure 9-42(a) shows, for example, that when two antennas spaced one-half wavelength apart are fed in phase, there is destructive interference along the array axis of the waves sent by the two antennas, since the propagation time delay represents a 180° phase shift between the two radiated waves. Perpendicular to the array, however, the waves arrive in phase and there is a constructive interference. Between these two extremes there is only partial addition.

If the antenna elements in an array each have an identical pattern, the resultant pattern can be obtained by multiplying the array pattern (assuming isotropic antennas with the same phases as in the array) by the individual antenna radiation pattern or unit pattern. This can best be seen by an example.

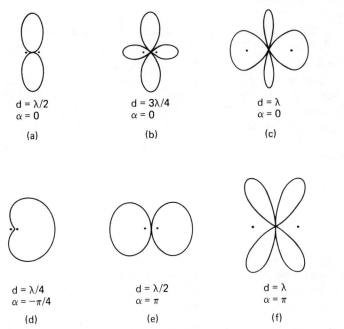

$d = \lambda/2$
$\alpha = 0$

(a)

$d = 3\lambda/4$
$\alpha = 0$

(b)

$d = \lambda$
$\alpha = 0$

(c)

$d = \lambda/4$
$\alpha = -\pi/4$

(d)

$d = \lambda/2$
$\alpha = \pi$

(e)

$d = \lambda$
$\alpha = \pi$

(f)

FIG. 9-42 Radiation patterns of two isotropic antennas when fed with equal currents at the phasings indicated.

EXAMPLE 9-6

Find the vertical radiation pattern in the plane shown (Fig. 9-43) of a horizontal $\lambda/2$ dipole located a height of $\lambda/4$ above a ground plane.

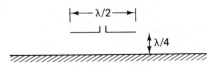

FIG. 9-43 A $\lambda/2$ dipole above a ground plane.

Solution: By employing the method of images, the ground plane can be replaced by a 180° out-of-phase antenna $\lambda/4$ below the ground-plane surface, as shown in Fig. 9-44.

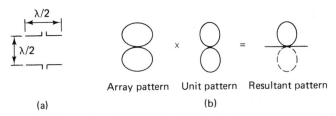

Array pattern   Unit pattern   Resultant pattern

(a)

(b)

FIG. 9-44 (a) Equivalent electrical circuit of a horizontal dipole antenna above the earth; (b) vertical radiation patterns obtained by using multiplication of patterns.

In order to obtain even greater directivities, much larger arrays are often employed. A very common type of array is the uniform linear array, where a number of elements are spaced equally along a straight line. These radiation patterns are somewhat idealized, since induced voltages and currents between the antennas in the array are neglected. The mutual coupling will affect the pattern of the individual antenna and thus of the resultant array pattern. In addition, the input impedance to each antenna depends upon the degree of coupling with all the neighboring antennas. Mutual coupling is greater for closely spaced elements and also stronger if the beam is directed along the array axis. These effects are usually measured under operating conditions.

9-15 TYPES OF ANTENNAS

As there are many types of antennas, with each having its own design criteria, we will merely list some antennas that are of current use (Table 9-1). For a more complete compendium of antenna design data and principles, one is referred to the *Antenna Engineering Handbook* edited by Jasik.[2] The information provided is only approximate, as the various parameters depend very much upon the design used.

9-16 MASTER ANTENNA TELEVISION SYSTEM

Let us apply some of the earlier theory to a simplified, yet somewhat typical *master antenna television system* (MATV). We shall, however, first obtain an estimation of the power power available at the terminals of a receiving antenna 9 m above ground, having a gain of about 9 dB, when receiving a signal from a TV transmitting antenna located 250 m above ground. Assume that 50 kW of power is transmitted on Channel 13 (210–216 MHZ) over the maximum line of sight distance, assuming a round, smooth earth. A typical transmitting antenna used in TV is the broadband dipole or superturnstile antenna shown in Fig. 9-45.

The two elements are fed 90° out of phase to obtain an omnidirectional pattern in the horizontal plane. The resulting electric field is horizontally polarized. The opening slots tend to compensate for the variations in impedance of the dipole over the bandwidth, resulting in an input impedance of about 75 Ω. By stacking these antennas, the resultant pattern can be altered, to give the desired radiation pattern. For our purposes we shall assume a dipole gain of 1.64. To obtain the maximum line-of-sight distance of a single antenna shown in Fig. 9-46, consider an antenna of height h located at the bulge center of Fig. 8-25, where $d = d_1 = d_2$. From equation (8-12) the line-of-sight distance to the horizon will be

$$d = \sqrt{2h} = 1.4\sqrt{h} \qquad (9\text{-}40a)$$

---

[2]Henry Jasik, *Antenna Engineering Handbook* (McGraw-Hill Book Company, 1961).

TABLE 9-1

Name	Radiation pattern	Polarization	Impedance	Gain	Bandwidth
Log-periodic dipole array $$t = \frac{R_{n+1}}{R_n} = \frac{l_{n+1}}{l_n}$$	E plane (plane containing elements) H plane (1'r to E plane) for $\alpha = 70°$ $\tau = 0.89$	Linear	$R = 60\ \Omega$ VSWR over period < 1.6	6.5 dB	10:1 Depends upon length of shortest and longest elements
Pyramidal horn ϕ's $< 40°$ Optimum horn $$\cos \phi/2 = \frac{L/\lambda}{S + L/\lambda}$$ where $S_H = 0.4$ $S_E = 0.25$	Horizontal 3 dB beam width $= \dfrac{70\lambda°}{d_H}$ Vertical 3 dB beam width $= \dfrac{56\lambda°}{d_E}$	Linear	VSWR < 2	$\dfrac{7.5\, d_E\, d_H}{\lambda^2}$	2:1

TABLE 9-1 (Continued)

Name	Radiation pattern	Polarization	Impedance	Gain	Bandwidth
Dielectric rod $l \approx 4\lambda^\circ$ for $\epsilon_r = 2.56$ $\dfrac{d_{min}}{\lambda}^\circ \approx 0.23$ $\dfrac{d_{max}}{d_{min}} \approx 1.6$	3 dB beamwidth $55\sqrt{\dfrac{\lambda_0^\circ}{l}}$	Linear	VSWR ≈ 1.5	$7\dfrac{l}{\lambda}$	10%
Wire trapezoidal tooth log periodic Period $= \ln \dfrac{1}{\tau}$ $\tau = \dfrac{R_{n+1}}{R_n}$	Vertical 3 dB beamwidth $\approx 70^\circ$ Horizontal 3 dB beamwidth $\approx 100^\circ$ for $\alpha = 60^\circ$ $\tau = 0.6$ $\psi = 35^\circ$	Linear	$R \approx 110\,\Omega$ VSWR over period <3	≈ 6 dB	10 : 1 Depends upon length of shortest and longest elements

TABLE 9-1 (Continued)

Name	Radiation pattern	Polarization	Impedance	Gain	Bandwidth
Helical $S \approx \lambda/4$ $\pi D \approx \lambda$ $\alpha \approx 14°$ No. of turns (N) > 3	3 dB beamwidth $$\frac{52\lambda}{\pi D}\sqrt{\frac{\lambda}{NS}}$$	Approximately circular	$R \approx 140\ \dfrac{\pi D}{\lambda}$ $X \approx 0$	$$\frac{15\ NS\ (\pi D)^2}{\lambda^3}$$	1.8 to 1
3 Element Yagi $d_r = \lambda/4$ $d_n = 0.2\lambda$ $l_r = \lambda/2$ $l = \lambda/2$ $l_d = 0.45\lambda$		Linear	$30 + j60$ May approach 50 Ω with larger array	≈ 7 dB Gains may approach 16 dB with larger array	10% Reduced BW with larger array

TABLE 9-1 (Continued)

Name	Radiation pattern	Polarization	Impedance	Gain	Bandwidth
Microwave parabolic dish	For uniformly illuminated aperture 3 dB beamwidth = $\dfrac{70\lambda^\circ}{D}$ 3 dB beamwidth between first nulls $\dfrac{149\lambda^\circ}{D}$	Depends on feed antenna	Depends on feed antenna	$\dfrac{K\pi^2 D^2}{\lambda^2}$ K is an efficiency constant, around 0.55 for most horn-fed antennas	Depends on feed antenna

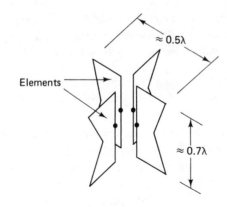

● Feed points

FIG. 9-45 Superturnstile antenna.

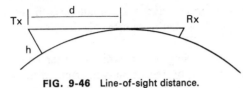

FIG. 9-46 Line-of-sight distance.

(where h is in feet and d is in miles) or in the metric system,

$$d = 4.1\sqrt{h} \qquad \text{(9-40b)}$$

(where h is in meters and d is in kilometers). Both of these equations take into account the slight bending of the rays in a standard atmosphere.

For an antenna height of 250 m (800 ft), a horizon distance of $4.1\sqrt{250}$ = 65 km is achieved. If the receiving antenna is elevated to 9 m, the line of sight will be increased by $4.1\sqrt{9}$ or 12 km, making a total range of about 77 km ($\approx$50 mi). The wavelength at Channel 13 is

$$\lambda = \frac{3 \times 10^8}{213 \times 10^6} = 1.4 \text{ m}$$

Assuming 100% efficient antennas and the maximum line-of-sight distance of 77 km, the power at the terminals of the receiving antenna can be obtained by substituting into equation (8-7), where the current in the transmitting antenna can be approximated by

$$I = \sqrt{\frac{P_t}{R_{\text{rad}}}} = \sqrt{\frac{50 \times 10^3}{75}} = 25.8 \text{ A}$$

$|E|$ at the receiving antenna $= \dfrac{240\pi \times 25.8 \times 250 \times 9}{1.4 \times (77 \times 10^3)^2}$

$$= 5.27 \text{ mV/m}$$

$\mathcal{P}$ at the receiving antenna $= \dfrac{E^2}{\eta} = 0.074 \ \mu\text{W/m}^2$

Since the receiving antenna gain = antilog $\frac{9}{10}$ = 8, the power at the rceiving terminals is $8 \times 0.074 \mu = 0.6 \mu$W. In actual practice the signal strength may vary considerably from this value, owing to such factors as:

1. The transmitting antenna may have directional characteristics which can result in either increased or decreased field strength at the receiving antenna.
2. The terrain may be rough, treed, have some water coverage, and so on, which may greatly affect the radiation field.
3. There may be obstacles such as buildings or other structures in the transmission path, or the receiving station may be in a valley or on a hill, greatly affecting the received signal. In weak-signal areas it may be best to monitor the signal strength with a field-strength meter or by employing a portable TV set.
4. The receiving antenna may not be pointed in the optimum direction if several channels are received on a single broadband antenna. Directional rotators can be installed to reduce this difficulty.

Let us now get on with the design of a typical master antenna television system (MATV), whereby a signal upon reception is distributed to various television sets. This avoids locating many separate antennas on, for instance, the top of a large apartment building.

The MATV system can be broken down into two main sections, the head and the distribution system.

Head End

The *head end* consists of the antenna, preamplifier, traps, filter, attenuator, and antenna mixer. A typical head end is shown in Fig. 9-47. Each component in the head end is briefly described in the same sequence as followed by the signal.

Receiving antenna This antenna is typically a 9-dB horizontally polarized yagi antenna used to capture as much of the signal as possible and to provide rejection of other local channels. Single-channel or narrow-band yagis may have a gain of a few dB above this, whereas a broad-band yagi has a gain somewhat lower. Physical size limits the number of elements in the antenna to about 10.

On average terrain, an improvement of about 6 dB in signal strength in the VHF band can be obtained by doubling the antenna height. It may be even better in the UHF regions. For even increased gain, the stacking of antennas can be employed, doubling the number of antennas doubles the gain.

A stack of four antennas provides from 5 to 6 dB gain improvement over a single antenna. In addition, stacking antennas provides for the opportunity to locate the nulls in the resulting pattern to be so positioned as to create maximum

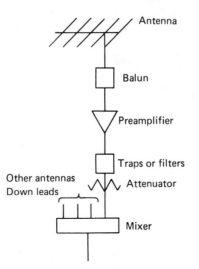

FIG. 9-47   Head end.

attenuation for undesired signals. Stagger stacking can also be used to improve front-to-back signals ratio up to 20 dB (refer to Fig. 9-42d). When cutting the coaxial line for the $\lambda/4$ difference, one must keep in mind the reduced wavelength in the dielectric as compared to that in air.

Baluns   These are used to convert from the 300 Ω balanced antenna terminals to the 75-Ω coaxial line. The opposite is done at the TV set.

Preamplifier   This is inserted to improve on the signal strength before it is lost in the noise due to attenuation. The noise figure of the preamplifier must be kept low, typically 5 dB. Preamplifier gains of 12–15 dB are readily available.

Coaxial Cable   Coaxial cable such as RG11 and RG59 produces less radiation interference than open-wire line and eliminates direct signal pickup. Losses increase with frequency and therefore is the greatest in Channel 13 of the VHF band.

Filters and Traps   These are used to eliminate undesired frequencies and to provide interference-free reception.

Attenuator   This is inserted on occasion to more-or-less equalize all signals coming into the antenna mixer. It will prevent overloading of the mixer in case of extremely large signals.

Mixer   The mixer is used to combine the signals from the various antennas before they reach the distribution amplifier. A loss of 6 dB can be expected.

Distribution System

The distribution system consists of a wideband amplifier, splitter(s), and tapoffs, as shown in Fig. 9-48.

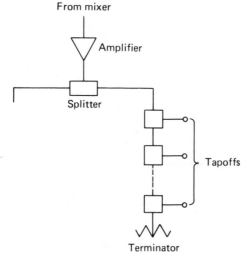

FIG. 9-48 Distribution system.

Amplifier This broad-band amplifier is inserted to overcome the losses caused by the distribution system. The output level should not be permitted to go into saturation as distortion and cross modulation begin to appear. In other words, the signal input plus amplifier gain should not exceed the output capacity of the amplifier. Gains can range up to 50 dB.

Splitters These separate or split the signal from the main line into several lines. Two-way splitters reduce the signal level by about 3.5 dB (0.5 dB of this is insertion loss), whereas four-way splitters reduce the signal levels another 3 dB to 6.5 dB.

Tapoffs These remove a portion of the signal from the line to the TV set while maintaining sufficient isolation to prevent interference between the sets

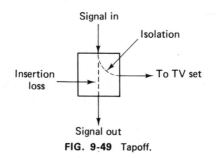

FIG. 9-49 Tapoff.

(Fig. 9-49). A minimum of 12 dB is recommended. The inherent loss for the main signal continuing on through the tapoff is called *insertion loss*.

Terminators This provides a good match at the transmission-line end to prevent reflections or ghosting.

In MATV work the reference chosen is 1000 μV across 75 Ω of impedance. This term is labeled as

$$\text{dBmV} = 20 \log \frac{V}{1000 \, \mu} \tag{9-41}$$

where V is the voltage in the system. For example, 1000 μV represents 0 dBmV and 3200 μV is equivalent to 10 dBmV. The 1000-μV reference is chosen, as it is the minimum signal that will produce snow-free pictures in even the old TV sets.

Let us consider the design of a simple MATV system whose branch with the largest loss has a total run of 60 m. It consists of one two-way splitter and five tapoffs, each with an insertion loss of 0.7 dB. Assume a cable loss of 0.14 dB/m. The total loss from splitter to fifth tapoff will be:

$$\text{Cable loss 60 m} \times 0.14 = 8.4 \text{ dB}$$
$$\text{Splitter loss} = 3.5 \text{ dB}$$
$$\text{Insertion loss } 4 \times 0.7 = 2.8 \text{ dB}$$
$$\underline{\text{Isolation loss of fifth splitter} = 12 \text{ dB}}$$
$$\text{Total distribution loss} = 26.7 \text{ dB}$$

Assuming that we require 2000 μV or 6 dBmV for the signal to the last TV set, we will need $26.7 + 6 = 32.7$ dBmV at the amplifier output. With a 25-dB gain amplifier, our signal at the preamplifier output must be 13.7 dBmV (6 dB loss in the mixer). With a 12-dB preamplifier, a minimum signal of 1.7 dBmV must be obtained from the antenna. This is just met by the 0.6 μW or, equivalently,

$$10 \log \frac{0.6 \, \mu}{(1000 \, \mu)^2/75} \text{ dBmV} = 1.7 \text{ dBmV}$$

anticipated by the system considered earlier. The levels along the MATV distribution system are shown in Fig. 9-50.

In actual practice, many more tapoffs and longer runs may be experienced, thus requiring higher gain amplifiers.

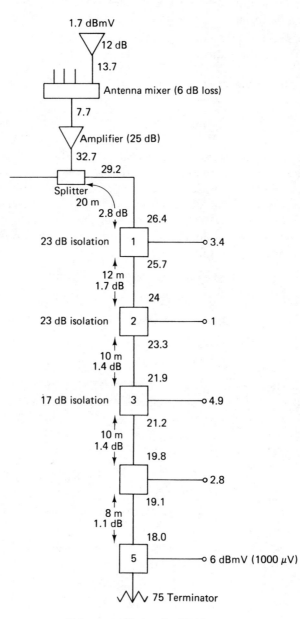

Units unspecified are in dBmV

FIG. 9-50 MATV system levels.

9-1. Radiation patterns are usually measured in the far-field region of an antenna. The minimum distance required to be in the far field is dependent upon the dimensions of the antenna in relation to the wavelength. The accepted formula for this distance is

$$r_{min} = \frac{2D^2}{\lambda}$$

where r_{min} = distance from the antenna
D = largest dimension of the antenna

A directional antenna has a maximum operating dimension of 50 m and operates at frequencies of 100 MHz and 2 GHz. Where does the far field begin at these two frequencies?

9-2. An antenna that radiates a total power of 80 W causes a maximum radiated field strength of 8 mV/m at a distance of 15 mi from the antenna. What is the directivity of this antenna (in dB)? If the antenna has an efficiency 95%, what is the maximum power gain of the antenna?

9-3. A center-fed half-wave dipole has a current of 12 A (rms) at its terminals. What is the field strength 1 mi from the dipole in a direction that is 30° from the dipole axis? What is the power density at this location?

9-4. Consider the following system in free space, with all components suitably matched.

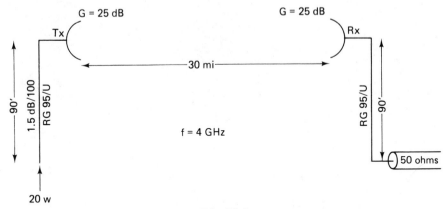

FIG. P9-4

(a) What is the size of this waveguide and what are it power-handling capabilities?
(b) Assuming 100% efficient antennas, what power is radiated by the transmitting antenna?
(c) What is the power density at the receiving antenna? (Assume proper alignment of antennas.)
(d) What power will be received by the 50 Ω line? What is the voltage across the line, and is it sufficiently strong for a sensitive receiver?

9-5. A radio station radiates a total power of 10 kW, having a gain of 6 dB. What is the electric field intensity at a distance of 100 km, expressed in volts per meter?

9-6. What is the loss (in dB) of radiated power density when moving from a 3-mi radial distance from a source to:
(a) A distance of 6 mi in the same radial direction?
(b) A distance of 12 mi in the same radial direction?

9-7.

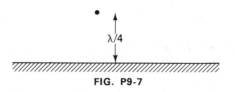

FIG. P9-7

A half-wavelength horizontal dipole is located at a height of $\lambda/4$ above a ground plane, as shown.
(a) Sketch the vertical radiation pattern in the plane shown.
(b) Determine the input resistance of the dipole over the ground plane.

9-8. Sketch the radiation pattern of two isotropic antennas spaced one-wavelength apart and having a 180° phase difference.

9-9. Sketch the radiation pattern of two half-wavelength dipoles fed 90° out of phase and having a spacing of $\lambda/4$ in the plane shown.

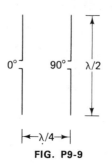

FIG. P9-9

9-10. (a) Sketch on semilog graph paper the free-space loss of two isotropic antennas spaced 30 mi apart over the frequency range 1–10 GHz.
(b) In view of your results in part (a), why can signals of reasonable strength still be received at the higher frequencies?

9-11. (a) A satellite transponder delivers 4 W of power to a 25-dB antenna; the feed losses amount to 0.5 dB. What is the EIRP of the satellite?
(b) On the down link the satellite transmits at a frequency of 4 GHz. Allowing for 0.3 dB atmospheric attenuation for rain, what will be the received power at the earth station in dBm if the gain of the receiving antenna is 58 dB? Assume a slant height to the satellite of 41,200 km (for 5° elevation). What is the received power (in μW)?
(c) What FSL (in dB) is experienced by the signal from the satellite [last term of equation (9-28)]?

appendix
A

THE EXPONENTIAL FUNCTION

The special irrational number 2.718, which is symbolized by e, is frequenctly encountered in electrical problems. When dealing with RC and RL networks, skin effect, transmission lines, and so on, the base of e is generally employed.

The exponential of the form e^x can be evaluated either from tables or on a calculator. The results of this expression is a real number.

EXAMPLE A-1

$$e^2 = 7.39.$$

EXAMPLE A-2

$$e^{-3} = 0.0498.$$

A complex number can also be written in the exponential form. In this case the exponent is imaginary and of the form $e^{j\theta}$.

Euler's theorem states that

$$e^{j\theta} = \cos\theta + j\sin\theta = 1\underline{/\theta} \qquad \text{(A-1)}$$

This indicates that $e^{j\theta}$ always has a magnitude of 1 ($\cos^2 \theta + \sin^2 \theta = 1$) and has a phase angle of θ. In this form, θ should be given in radians. This expression can be graphically illustrated as shown in Fig. A-1. If $e^{j\theta}$ is multiplied by a constant r, then

$$re^{j\theta} = r(\cos \theta + j \sin \theta) = r\underline{/\theta} \tag{A-2}$$

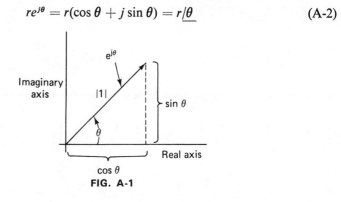

FIG. A-1

These various forms are usually given the names

exponential form when written as $re^{j\theta}$

rectangular form when written as $r \cos \theta + jr \sin \theta$

polar form when written as $r\underline{/\theta}$

A few examples will now be given.

EXAMPLE A-3

Express $10\,e^{j\pi/2}$ in rectangular form:

Solution:

$$10e^{j\pi/2} = 10\left(\cos \frac{\pi}{2} + j \sin \frac{\pi}{2}\right)$$

$$= j10$$

This is graphically represented in Fig. A-2.

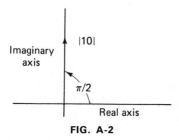

FIG. A-2

EXAMPLE A-4

Express $4e^{-j\pi/4}$ in polar and rectangular form.

Solution: This expression is represented graphically in Fig. A-3.

$$4e^{-j\pi/4} = 4\underline{/-\pi/4} = 4\underline{/-45°} \quad \text{(polar form)}$$

$$4\underline{/-45} = 4\cos 45 - j4\sin 45$$

$$= 2.83 - j2.83 \quad \text{(rectangular form)}$$

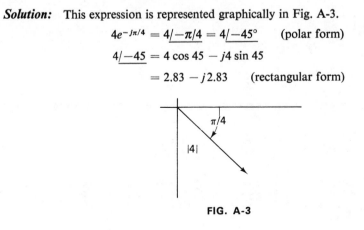

FIG. A-3

appendix
B

DERIVATION
OF THE SMITH CHART

The Smith chart is basically a graphical representation of the impedance one sees as one moves along a transmission line. Its derivation can be obtained by beginning with the impedance equation (3-51). Upon normalizing this equation (dividing through by Z_0) and substituting d for l, we obtain for the normalized impedance anywhere along the line.

$$\frac{Z}{Z_0} = \frac{e^{\gamma d} + \Gamma_R e^{-\gamma d}}{e^{\gamma d} - \Gamma_R e^{-\gamma d}} \tag{B-1}$$

Letting the load reflection coefficient (Γ_R) have a magnitude of $|\Gamma_R|$ and a phase angle of ϕ(i.e., $\Gamma_R = |\Gamma_R| \underline{/\phi}$), the normalized impedance can be written as (after dividing both numerator and denominator by $e^{\gamma d}$)

$$\frac{Z}{Z_0} = \frac{1 + |\Gamma_R| e^{-2\alpha d} e^{-j(2\beta d - \phi)}}{1 - |\Gamma_R| e^{-2\alpha d} e^{-j(2\beta d - \phi)}} \tag{B-2}$$

where $\gamma = \alpha + j\beta$ has been substituted.

Let us consider initially a lossless transmission line where $\alpha = 0$. Then equation (B-2) can be written as

$$\frac{Z}{Z_0} = \frac{1 + |\Gamma_R| \underline{/\phi - 2\beta d}}{1 - |\Gamma_R| \underline{/\phi - 2\beta d}} \tag{B-3}$$

The numerator and denominator of the last equation is plotted graphically in Fig. B-1. It can be noted from this diagram that when the numerator is maximum, this occurs when the term $|\Gamma_R|\underline{/\phi - 2\beta d}$ is in phase and collinear with 1 (unity), the denominator is a minimum. This particular point, the normalized impedance, is a maximum and equal to

$$\left(\frac{Z}{Z_0}\right)_{max} = \frac{1 + |\Gamma_R|}{1 - |\Gamma_R|} \tag{B-4}$$

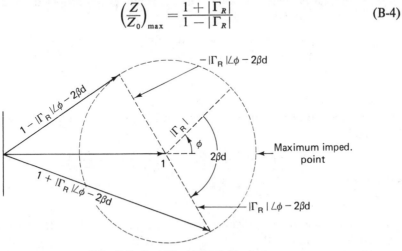

FIG. B-1 Evolution of the Smith chart.

which is a positive and a real quantity since the maximum value for the reflection coefficient is 1. From equation (3-75) it is seen that equation (B-4) is equal to the VSWR. We can conclude that the *maximum normalized impedance* on a lossless line is equivalent to the VSWR, which is a real quantity.

$$\text{VSWR} = \left(\frac{Z}{Z_0}\right)_{max} \tag{B-5}$$

Since the Smith chart consists of resistance and reactance loci, let

$$\frac{Z}{Z_0} = r + jx \tag{B-6}$$

where r = normalized resistance
x = normalized reactance

For simplification, also let

$$|\Gamma_R|\underline{/\phi - 2\beta d} = u + jv \tag{B-7}$$

The right side of equation (B-7) is the rectangular form of the left side of the equation, which is in polar form (see Appendix A). Thus,

$$\frac{Z}{Z_0} = r + jx = \frac{1 + u + jv}{1 - u - jv} = \frac{1 - u^2 - v^2 - 2jv}{(1 - u)^2 + v^2} \tag{B-8}$$

The last term in equation (B-8) is derived by multiplying both the denominator and numerator of the previous term by the conjugate of the denominator $(1 - u + jv)$.

Equating the real and imaginary parts, we obtain

$$r(1 - 2u + u^2 + v^2) = 1 - u^2 - v^2 \tag{B-9}$$

$$x(1 - 2u + u^2 + v^2) = 2v \tag{B-10}$$

Since the first equation involves r only, we can obtain the constant resistance loci from it. Similarly, one can obtain the constant reactance loci from the second equation. From equation (B-9),

$$r - 2ru + ru^2 + rv^2 = 1 - u^2 - v^2$$

$$u^2(r + 1) - 2ru + v^2(1 + r) = 1 - r$$

$$u^2 - \frac{2r}{r + 1}u + v^2 = \frac{1 - r}{1 + r}$$

This is an equation of a circle. To determine its center and radius, its square must be completed.

$$\left(u - \frac{r}{r + 1}\right)^2 + v^2 = \frac{1 - r}{1 + r} + \left(\frac{r}{r + 1}\right)^2$$

$$\left(u - \frac{r}{r + 1}\right)^2 + v^2 = \frac{1}{(r + 1)^2} \tag{B-11}$$

This is the equation for a circle having a radius of $1/(r + 1)$ and centered at $u = r/(r + 1)$ and $v = 0$.

Table B-1 gives values for the circle radius and its center for various values of the normalized resistance. Figure B-2 is a graphical plot of equation (B-11) using the values given in the table. From equation (B-10),

$$x - 2xu + xu^2 + xv^2 = 2v$$

$$u^2 - 2u + v^2 - \frac{2v}{x} = -1$$

TABLE B-1   Values for the Circle Radius and Its Center

Normalized resistance, r	Radius	Centered at: v	Centered at: u
0	1	0	0
$\frac{1}{2}$	$\frac{2}{3}$	0	$\frac{1}{3}$
1	$\frac{1}{2}$	0	$\frac{1}{2}$
2	$\frac{1}{3}$	0	$\frac{2}{3}$

Completing its square, we obtain

$$(u - 1)^2 + \left(v - \frac{1}{x}\right)^2 = -1 + 1 + \frac{1}{x^2} = \left(\frac{1}{x}\right)^2 \tag{B-12}$$

This equation represents a circle, having a radius $1/x$ and centered at $u = 1$, $v = 1/x$.

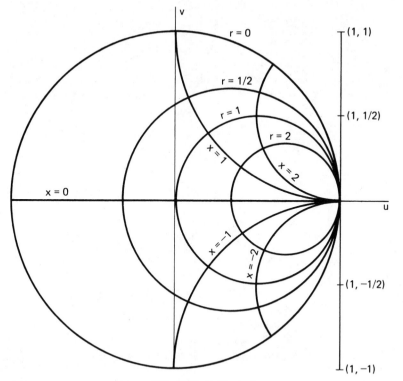

FIG. B-2 Smith chart.

For various values of x, the radius of the corresponding circle and its center are given in Table B-2. Figure B-2 also plots these values.

TABLE B-2 Values for the Circle Radius and Its Center

Normalized reactance, x	Radius	Centered at: v	u
0	∞	∞	1
± 1	1	± 1	1
± 2	$\frac{1}{2}$	$\pm \frac{1}{2}$	1

Since the reflection coefficient is also expressed in terms of u and v [equation (B-7)], it can also be plotted on the Smith chart. This is done in Fig. B-3. As can be seen from the figure, as one moves toward the generator on a transmission line (increasing d), one rotates in a clockwise direction on the Smith chart (an angular distance of $2\beta d$ radians). Most Smith charts indicate this distance in wavelengths around the periphery of the chart (see Fig. 4-1).

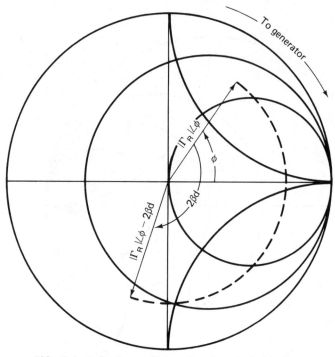

FIG. B-3 Reflection coefficient plotted on the Smith chart.

In the case of a lossy line, the amplitude of the reflection coefficient also changes [see equation (4-3)] by the factor $e^{-2\alpha d}$:

$$|\Gamma| = |\Gamma_R|e^{-2\alpha d} \tag{B-13}$$

The amplitude of the reflection coefficient decreases as one advances toward the generator.

appendix
C

$TE_{m,n}$ AND $TM_{m,n}$ FIELDS IN A RECTANGULAR WAVEGUIDE

$TE_{m,n}$ *Fields* $(E_z = 0)$

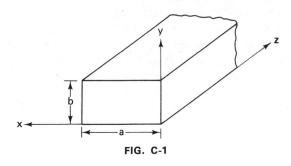

FIG. C-1

$$E_x = j\frac{60\pi ncfH_0}{b\epsilon_r f_c^2} \cos\frac{m\pi x}{a} \sin\frac{n\pi y}{b} e^{-j\beta}g^z \qquad (C\text{-}1a)$$

$$E_y = -j\frac{60\pi mcfH_0}{a\epsilon_r f_c^2} \sin\frac{m\pi x}{a} \cos\frac{n\pi y}{b} e^{-j\beta}g^z \qquad (C\text{-}1b)$$

$$E_z = 0 \qquad (C\text{-}1c)$$

$$H_x = j\frac{mc\sqrt{f^2 - f_c^2}H_0}{2a\sqrt{\mu_r\epsilon_r}f_c^2} \sin\frac{m\pi x}{a} \cos\frac{n\pi y}{b} e^{-j\beta}g^z \qquad (C\text{-}1d)$$

$$H_y = j \frac{nc\sqrt{f^2 - f_c^2}\, H_0}{2b\sqrt{\mu_r \epsilon_r}\, f_c^2} \cos \frac{m\pi x}{a} \sin \frac{n\pi y}{b}\, e^{-j\beta g^z} \tag{C-1e}$$

$$H_z = H_0 \cos \frac{m\pi x}{a} \cos \frac{n\pi y}{b}\, e^{-j\beta g^z} \tag{C-1f}$$

TM$_{m,n}$ Fields (*H*$_z$ = 0)

$$E_x = \frac{-jmc\sqrt{f^2 - f_c^2}}{2a\sqrt{\mu_r \epsilon_r}\, f_c^2}\, E_0 \cos \frac{m\pi x}{a} \sin \frac{n\pi y}{b}\, e^{-j\beta g^z} \tag{C-2a}$$

$$E_y = \frac{-jnc\sqrt{f^2 - f_c^2}}{2b\sqrt{\mu_r \epsilon_r}\, f_c^2}\, E_0 \sin \frac{m\pi x}{a} \cos \frac{n\pi y}{b}\, e^{-j\beta g^z} \tag{C-2b}$$

$$E_z = E_0 \sin \frac{m\pi x}{a} \sin \frac{n\pi y}{b}\, e^{-j\beta g^z} \tag{C-2c}$$

$$H_x = \frac{jncf}{240\pi \mu_r b f_c^2}\, E_0 \sin \frac{m\pi x}{a} \cos \frac{n\pi y}{b}\, e^{-j\beta g^z} \tag{C-2d}$$

$$H_y = \frac{-jmcf}{240\pi \mu_r a f_c^2}\, E_0 \cos \frac{m\pi x}{a} \sin \frac{n\pi y}{b}\, e^{-j\beta g^z} \tag{C-2e}$$

$$H_z = 0 \tag{C-2f}$$

where

$$f_c = f_{c(m,n)} = \frac{c}{2\sqrt{\mu_r \epsilon_r}} \sqrt{\left(\frac{m}{a}\right)^2 + \left(\frac{n}{b}\right)^2} \tag{C-3}$$

The term f_c is the cutoff frequency. When operating at a frequency above f_c, a wave propagates down the waveguide and the mode is called a *propagating mode*. When operating below the cutoff frequency, the field decays exponentially and there is no wave propagation. In this case, the mode is called a *nonpropagating mode* or an *evanescent mode*. H_0 and E_0 are arbitrary amplitude constants in the equations which depend upon the source supplying the energy to the waveguide.

appendix
D

WAVEGUIDE
DESIGNATIONS

TABLE D-1 Waveguide Designations

Frequency range TE$_{10}$-mode 153-IEC* GHz	Cut-off freq. TE$_{10}$ Mode GHz	Waveguide designation					Waveguide inner cross-section 153-IEC*			
		153-IEC*	British stand.	Retma	JAN RG-/U brass	JAN RG-/U alum.	Band prefix	Width mm	Height mm	Tolerance on width and height ±
1.14– 1.73	0.908	R 14	WG 6	WR 650	69	103	L	165.10	82.55	0.33
1.45– 2.20	1.158	R 18	WG 7	WR 510	—	105	D	129.54	64.77	0.26
1.72– 2.61	1.375	R 22	WG 8	WR 430	104	—	—	109.22	54.61	0.22
2.17– 3.30	1.737	R 26	WG 9A	WR 340	112	113	—	86.36	43.18	0.17
2.60– 3.95	2.080	R 32	WG 10	WR 284	48	75	S	72.14	34.04	0.14
3.22– 4.90	2.579	R 40	WG 11A	WR 229	—	—	A	58.17	29.083	0.12
3.94– 5.99	3.155	R 48	WG 12	WR 187	49	95	G	47.55	22.149	0.095
4.64– 7.05	3.714	R 58	WG 13	WR 159	—	—	C	40.39	20.193	0.081
5.38– 8.17	4.285	R 70	WG 14	WR 137	50	106	J	34.85	15.799	0.070
6.57– 9.99	5.260	R 84	WG 15	WR 112	51	68	H	28.499	12.624	0.057
7.00 –11.00	5.790	—	—	WR 102	—	320	T	25.90	12.95	0.125
8.2 – 12.5	6.560	R 100	WG 16	WR 90	52	67	X	22.860	10.160	0.046
9.84– 15.0	7.873	R 120	WG 17	WR 75	—	—	M	19.050	9.525	0.038
11.9 – 18.0	9.490	R 140	WG 18	WR 62	91	107	P	15.799	7.899	0.031
14.5 – 22.0	11.578	R 180	WG 19	WR 51	—	—	—	12.954	6.477	0.026
17.6 – 26.7	14.080	R 220	WG 20	WR 42	53	121	—	10.668	4.318	0.021
21.7 – 33.0	17.368	R 260	WG 21	WR 34	—	—	—	8.636	4.318	0.020
26.4 – 40.0	21.100	R 320	WG 22	WR 28	96	—	Q	7.112	3.556	0.020
32.9 – 50.1	26.350	R 400	WG 23	WR 22	97	—	—	5.690	2.845	0.020
39.2 – 59.6	31.410	R 500	WG 24	WR 19	98	—	—	4.775	2.388	0.020
49.8 – 75.8	39.900	R 620	WG 25	WR 15	99	—	—	3.759	1.880	0.020
60.5 – 91.9	48.400	R 740	WG 26	WR 12	—	—	—	3.099	1.549	0.020
73.8 –112.0	59.050	R 900	WG 27	WR 10	—	—	—	2.540	1.270	0.020
92.2 –140.0	73.840	R 1200	WG 28	WR 8	138	—	—	2.032	1.016	0.020
114.0 –173.0	90.845	R 1400	WG 29	WR 7	136	—	—	1.651	0.826	—

TABLE D-1 (Continued)

Frequency range TE$_{10}$-mode 153-IEC* GHz	Cut-off freq. TE$_{10}$ Mode GHz	Waveguide outer cross-section 153-IEC*			Attenuation in dB/m for copper waveguide 153-IEC*			Theoretical C.W. power rating** lowest to highest frequency MW
		Width mm	Height mm	Tolerance on width and height ±	Frequency GHz	Theoretical value	Maximum value	
1.14– 1.73	0.908	169.16	86.61	0.20	1.36	0.00522	0.007	12.0 –17.0
1.45– 2.20	1.158	133.60	68.83	0.20	1.74	0.00749	0.010	7.5 –11.0
1.72– 2.61	1.375	113.28	58.67	0.20	2.06	0.00970	0.013	5.2 – 7.5
2.17– 3.30	1.737	90.42	47.24	0.17	2.61	0.0138	0.018	3.4 – 4.8
2.60– 3.95	2.080	76.20	38.10	0.14	3.12	0.0189	0.025	2.2 – 3.2
3.22– 4.90	2.579	61.42	32.33	0.12	3.87	0.0249	0.032	1.6 – 2.2
3.94– 5.99	3.155	50.80	25.40	0.095	4.73	0.0355	0.046	0.94 – 1.32
4.64– 7.05	3.714	43.64	23.44	0.081	5.57	0.0431	0.056	0.79 – 1.0
5.38– 8.17	4.285	38.10	19.05	0.070	6.46	0.0576	0.075	0.56 – 0.71
6.57– 9.99	5.260	31.75	15.88	0.057	7.89	0.0794	0.103	0.35 – 0.46
7.00– 11.00	5.790	29.16	16.21	0.125	—	—	—	0.33 – 0.43
8.2 – 12.5	6.560	25.40	12.70	0.05	9.84	0.110	0.143	0.20 – 0.29
9.84– 15.0	7.873	21.59	12.06	0.05	11.8	0.133	—	0.17 – 0.23
11.9 – 18.0	9.490	17.83	9.93	0.05	14.2	0.176	—	0.12 – 0.16
14.5 – 22.0	11.578	14.99	8.51	0.05	17.4	0.238	—	0.080 – 0.107
17.6 – 26.7	14.080	12.70	6.35	0.05	21.1	0.370	—	0.043 – 0.058
21.7 – 33.0	17.368	10.67	6.35	0.05	26.1	0.435	—	0.034 – 0.048
26.4 – 40.0	21.100	9.14	5.59	0.05	31.6	0.583	—	0.022 – 0.031
32.9 – 50.1	26.350	7.72	4.88	0.05	39.5	0.815	—	0.014 – 0.020
39.2 – 59.6	31.410	6.81	4.42	0.05	47.1	1.060	—	0.011 – 0.015
49.8 – 75.8	39.900	5.79	3.91	0.05	59.9	1.52	—	0.0063– 0.0090
60.5 – 91.9	48.400	5.13	3.58	0.05	72.6	2.03	—	0.0042– 0.0060
73.8 –112.0	59.050	4.57	3.30	0.05	88.6	2.74	—	0.0030– 0.0041
92.2 –140.0	73.840	4.06	3.05	0.05	111.0	3.82	—	0.0018– 0.0026
114.0 –173.0	90.845	—	—		136.3	5.21	—	0.0012– 0.0017

*IEC recommendations are obtainable from: Central Office of the International Electrotechnical Commission, 1, rue de Varembé, Geneva, Switzerland

**Based on breakdown of air of 15,000 volts per cm (safety factor of approximately 2 at sea level).

Source: Philips Electronic Industries Ltd.

INDEX